Complexity Management in Engineering Design – a Primer

Maik Maurer

Complexity Management in Engineering Design – a Primer

Maik Maurer
LSt f. Produktentwicklung
Technische Universität München
Garching, Bayern
Germany

ISBN 978-3-662-53447-2   ISBN 978-3-662-53448-9   (eBook)
DOI 10.1007/978-3-662-53448-9

Library of Congress Control Number: 2016963068

Printed on acid-free paper

This Springer Vieweg imprint is published by Springer Nature
The registered company is Springer-Verlag GmbH Heidelberg
The registered company address is: Heidelberger Platz 3, 14197 Berlin, Germany

Foreword

Product Development is struggling with an increasing complexity due to requirements, regulations, technologies, supporting digital tools, etc. The handling of an increasing quantity of data is supported by PDM and PLM systems; requirements are managed with the help of databases. However, design engineers have hardly any support for obtaining a transparent overview in a holistic way, including products, processes, data, organization, means, etc.

The starting point of Dr. Maik Maurer's research in the field of complex systems was the vision of individualized physical products based on mass production. Out of the large number of questions arising, he focused on more transparency and understanding of the interdependencies that lead to difficulties in generating solutions. Based on his research questions related to this vision, he started to evaluate the possibilities of modeling elements and interdependencies in and between different domains: requirements, products, and processes etc., based on mathematical models. Methods as well as graph theory deliver the required input. Existing methods and newly developed ones have been linked together to build a framework for the management of structural complexity. Interdependencies and patterns out of these interdependencies across domains are the basis for the analysis.

Strategies and methods for data acquisition and quality assurance of the structural data, together with possibilities for visualization by strength-based graphs, complete the approach.

Several research projects funded by the German Science Foundation as well as a technology-transfer project with applications in different industries and organizations formed an excellent basis for improvement and further development, as well as an evaluation of the methodological framework. The associates in the technology transfer range from some well-known international corporate groups to medium-sized companies. The framework developed also formed one of the building blocks of a start-up consulting firm.

Because of the strong link between our research and the education of degree students, an increasing number of theses on the subject were written. A specific lecture for master's-degree students about structural complexity started with about 30 students and continuously grew to well over 100 per year. An intensive practical course in systems engineering complemented this lecture for master's-degree students.

An international summer school on systems engineering initiated by Maik Maurer started in Germany; it since became Europe-wide, and about 20 doctoral students participated per year.

Based on these activities and experiences, Maik Maurer wrote this book as his habilitation thesis, entitled "Complexity management in engineering design—a primer." Complexity, and the lack of transparency due to it, is one of the most demanding aspects of engineering design and beyond. Maik Maurer addressed this in his thesis, based on a unique overview of the historical development of complexity management. He presents a classification of approaches and develops a management framework for handling complexity, especially in engineering design.

Based on a number of years of research, and its successful transfer to industry, Maik Maurer presents an impressive work for teaching that is relevant for both students and practitioners.

Garching   Udo Lindemann
August 2016

Acknowledgement

This work is the result of extensive study of the topic of complexity from various perspectives in research, teaching, and industrial applications in an always positive and inspiring work environment. Without the support, suggestions, and contributions of many people, I would not have been able to create this work in its existing form.

First of all, I would like to thank Prof. Udo Lindemann for his many years of support and mentorship in a variety of research, teaching, and industrial projects. His support enabled me to attack even unconventional and risky projects with the greatest confidence and commitment.

I would also like to express my highest gratitude to all of my colleagues at the Institute of Product Development for their cooperation. It has been the key to the successful application for and conducting of research projects and for designing, conducting, and continuously improving lecture courses and programs.

Special thanks also go to all the students I was allowed to teach and supervise during various courses and theses. Their critical questions, proposed alternative solutions, and well-considered feedback to my lecturing helped me to develop and improve as a teacher. The need to not only comprehend a topic myself but also to be able to transfer this knowledge to others has deepened my understanding of matters substantially.

Finally, I would particularly like to thank my family for always supporting me throughout the time of creating this thesis and for their understanding of my long evenings or weekends spent in the office.

Stoneham, MA
February 2016

Maik Maurer

Contents

Initial Thoughts

In recent times it seems that the term complexity is omnipresent and almost always gets associated with negative effects and consequences. Complexity means high efforts to successfully manage a project, many resources required for execution, high (and very often overstretched) budgets and tremendous risks of failure. Therefore, it is understandable that people tend to avoid complexity—as it seems to mean trouble.

One implication of complexity impacting more and more areas of job-related and private lives is that people are often open to simplification approaches. Simplify Your Life was the title of a bestseller; several similar phrases appeared as book titles in the last decade. While purification from useless complexity seems to be promising, those concepts do not represent a silver bullet to the realities in engineering.

One point that often gets neglected is: Complexity works! Nature, animals and humans are highly complex systems. And while we do only understand parts of the functionality and interactions of these systems, we have to state that they work impressively well. And compared to the epoch of industrialization and the product lifetime of technical systems, highly complex natural systems seem to outperform anything we design.

Simplification as a cure to the challenges of complexity also neglects another important aspect: the amount of functionalities of a system seems to be related to the degree of its complexity. Or in other words: a large range of functionality cannot result from a simple mechanism with only a few components. That is to say, a replica of the human brain will not be produced from a handful of Lego blocks.

As much as unnecessary complexity should be avoided, there is much useful complexity that is essential for society, technology and progress in general. Of course, we will be able to simplify a specific function the more we know about it and the more we can control it. But to integrate more functionality into the same product with less complexity? Can modern cars with all their driver assistance systems, engine management systems, diagnosis systems, etc. be less complex than the Ford Model T? There is no way.

© Springer-Verlag GmbH Germany 2017
M. Maurer, *Complexity Management in Engineering Design – a Primer*,
DOI 10.1007/978-3-662-53448-9_1

So, engineers have to deal with complexity and it is not an option to bury one's head in the sand. Therefore it is useful to know about the origins, history and common approaches towards complexity management. The following chapters will provide an overview of complexity and the engineering approaches for dealing with it from different perspectives.

This thesis shall mediate that complexity is a natural characteristic of many products, processes and systems. Avoiding or reducing this complexity would often degrade functionality and is not a viable option. The thesis shall give an overview of fundamentals and state of applications, and provide a general framework for tackling complex challenges in engineering.

Industrial applicants should benefit from the experiences resulting from a long historical development that involves several scientific as well as industrial disciplines.

Scientific researchers should especially benefit from the classification of complexity management approaches shown in this thesis. Overlaps of research fields as well as "blank spots" provide an indicator for research efforts required in the future in order to better deal with inevitable complex systems.

If any more motivation is required for taking up the complexity challenge: The author is convinced that it is not the ability of simplification, but the ability to efficiently build, manage and interact with complex systems that is the key competency for the future of engineering.

The Omnipresence of Complexity

Today the term complexity suffers from inflationary use, having become a buzzword that is applied without really knowing about its specific meaning. Following the news makes it seem like "complexity increases" in almost all areas and "complexity is the most important challenge" of the future. Sometimes, descriptions of "complexities" appear Apparently, the plural gets applied in the absence of a comparative degree and not to express the occurrence of different types of complexity. The term complexities gets used for overbidding "normal complexity". When asking the authors of such statements for a definition of the fundamental challenges, this often remains unclear.

Declaring a problem as being complex is often based on insufficient knowledge about the situation. Or in other words: If one characterizes a situation or question as complex, he often means that the original cause of an undesired effect is not transparent—and therefore cannot be treated.

Methods provide procedures and support systematic problem-solving. Numerous methods have been designed for managing complexity—focusing on different complexity origins, interacting with different types of system elements and pursuing different objectives. Just declaring a question as complex and selecting a method is not enough. The selection of a method requires knowledge about the complex problem and certainty about the desired objective. The dilemma is that if a "complex" problem means that it is not understood (not transparent), it is hardly possible to determine a specific objective. Thus, the acknowledgement of complexity cannot directly be followed up by selecting a method for its treatment, but requires measures for creating a better understanding first. When selecting and successfully applying a method, one must ensure that the purpose of the method fits the problem at hand. Systematic complexity management initially identifies underlying causes for the undesired effects called complexity, specifies the type of complexity and then determines suitable strategies and methods to be applied for solving the challenge.

© Springer-Verlag GmbH Germany 2017
M. Maurer, *Complexity Management in Engineering Design – a Primer*,
DOI 10.1007/978-3-662-53448-9_2

The following sections of this chapter briefly introduce the basics of complexity as an important engineering challenge, depict the structure and specify the target group of the thesis.

2.1   About the Basic Complexity Survival Skills for Engineers

As omnipresent as the existence (or at least the assumption) of complexity is, the strategies, methods and tools offered for its management are equally numerous. A somewhat radical strategy gained significant traction in the popular book market: Simplify Your Life has become a bestseller promoting the notion of overcoming everyday complexity by avoiding or even ignoring it [1]. While some of those approaches may work for managing an overloaded personal calendar, professional work problems like a complex production process flow should—obviously—not be managed by ignoring its existence.

Strategies of complexity management aim at curing the origins of complexity or help to mitigate its impacts. For example, Schuh and Schwenk laid out a straightforward procedure for handling the excessive creation of product variants [2]. If excessively numerous product variants represent the negative impact of complexity, tackling the creation of new variants focuses on the root cause. Baldwin and Clark describe how architectural component interdependencies can be controlled and optimized in order to minimize the product or product portfolio complexity in a modular or integral product structure [3]. If the effect of complexity is an unmanageable impact between interconnected components, then such an approach with reduced interfaces tackles the cause.

These two examples of counteracting complexity depict the major challenge: How do we determine the right method of complexity management if the origin of this complexity is not obvious? The methods of variant management mentioned before, as introduced by Schuh and Schwenk, can be very powerful—but not in the specific case if the complexity is emerging from the unmanageably large number of component interconnections within a single product.

So even if an engineer knows several approaches to handle complexity, selecting one without profound understanding of the problem is like grabbing an arbitrary tool from a toolbox only knowing that the car makes "some strange noise"—nobody would treat a problem this way. And if an engineer only knows about one approach towards complexity management, it could be a risk that once he is facing complexity in his daily work this becomes the tool of choice. Of course, such a narrowly focused procedure is doomed to fail eventually, as different situations, objectives and boundary conditions require different methodical approaches.

Thus, complexity often implies a lack of understanding in the problem domain, incomplete information and uncertainty. If the problem were understood and all required information were available, then the associated system could be modeled and various solution approaches be applied. This narrows down the challenge of complexity management: If confronted with a complex problem, we first need to understand about the origin of complexity before we can purposefully decide on a strategy of management.

As stated in the title of this chapter, complexity is omnipresent and we deal with it every day. It is not something new or extraordinary. Just the impressive amount of sensory input humans continuously receive from their different senses presents a complex challenge to manage—and we manage it—continually. However, managing the complexity emerging from artificial, technical systems seems to be a poor fit with our innate complexity management abilities. While humans are easily able to pick out relevant information from thousands of simultaneous and competing visual impressions, the same humans cannot reliably comprehend a network of processes containing a couple dozen elements. Educational management games based on system dynamics are often applied to illustrate how easily even experienced managers can destabilize (hypothetical) supply chains or entire enterprises [4, 5].

As the human ability to control natural complexity cannot be applied to technical challenges and the wide variety of technical complexity asks for specifically appropriate approaches, a sound understanding of complexity, its characteristics and specifications is essentially required. This is then the basis for investigating appropriate procedures and methods to guide and support engineers in solving unavoidable complex challenges.

Not only is (a specific kind of) complexity management a natural ability of humans, complexity management for non-natural systems has a surprisingly long history. More than 2000 years ago, Greek philosophers began tackling societal and governmental challenges, which were characterized by significant complexity. And from these times on, systems and their inherent complexity became described, analyzed and managed with many approaches—based and embedded into the specific time. It is an important insight that complexity and its management are not challenges of recent times only. However, one could get this false impression, as in the last few years complexity gets heavily used for explaining problems. In fact, complexity is often used as an excuse, such as when projects miss their objectives and when major accidents happen. But the historical development shows that complexity and its management underwent a long-lasting evolution—and every epoch had to tackle its complexity using the means available at the time.

Today, interacting with complexity is crucial in many fields, and also almost everywhere in engineering. For various reasons, knowledge about procedures, methods and tools are not equally distributed among the different engineering disciplines. These disciplines were founded at different times—e.g. software engineering emerged later than product development—and have different states of the art and use different resources. Understanding the links and commonalities between the disciplines can support a sound basis for identifying opportunities to transfer knowledge concerning complexity management.

This thesis shall contribute to gaining a better understanding of the phenomena aggregated in the term "complexity" in an engineering context. This understanding shall be reached by classifying complexity by relevant criteria, differentiating between complicated and complex challenges, investigating useful definitions of complexity and characterizing the impact resulting from complexity. In addition, well-established complexity management approaches like variant management, interface management and Pareto analysis are introduced, and their range of application and methodic background are described.

As the sole discovery of complexity does not clarify its origins in a system, this thesis shall further provide insight into where complexity emerges from in the engineering field. Knowing the root causes of observed complexity and being able to specify this complexity, these represent the basis for selecting appropriate methods and tools. For such applications, the thesis introduces a generic framework.

2.2   Who Should Read This Thesis?

The content of this thesis is partly based on the lecture "Complexity Management for industrial applications" (German: Komplexitätsmanagement für die industrielle Praxis), which the author held from 2009 to 2014 at the Technical University of Munich, Germany (Master's program in Mechanical Engineering) and content originating from business trainings the author conducted for several years with numerous companies.

Furthermore, findings from several academic research projects in the field of structural complexity management and knowledge management, which the author conducted from 2008 to 2014, have been integrated. And industrial insights gained from more than 20 consultancy projects and development of complexity management software over a period of 7 years also contribute to the basis for this thesis.

This thesis is aimed to be a supplemental source for students of engineering disciplines, who became aware of specific challenges of complexity and want to get a more general insight into the topic. Instead of teaching a single or some specific methods, this thesis should provide a big picture of complexity management, guiding students through the necessities, ideas, concepts and implementations. The thesis is meant to prevent an isolated view on complexity, either that complexity is harmful and should avoided at all means or that there is "the one solution" to manage complexity. Thus, this thesis should mediate an awareness of complexity as a natural, unavoidable characteristic that is necessary to control and manage.

This thesis is also meant to serve engineers in practice as a guide towards successful complexity management. It shall show engineers how to make complex issues transparent and enable them to identify the right methods and tools for their specific complexity challenges. By explaining the term complexity, its origins, history and the established methods to tackle it, the thesis gives engineers a framework at hand for identifying needs and possibilities in dealing with complexity as it appears in their day-to-day projects.

2.3   Structure of the Thesis

The thesis is partitioned into four main chapters numbered 3–6, which introduce the occurrence of complexity in engineering, the historical background of complexity management, a classification of approaches towards handling complexity in engineering disciplines and a framework for application.

Chapter 3 focuses on providing an understanding of the phenomena of complexity in engineering, distinguishing complexity from other challenges and providing an overview of common definitions. After introducing the challenge of complexity, the commonly applied approaches of their management in engineering are introduced.

Chapter 4 describes the historical background of complexity management, highlighting the important epochs, their key actors and their discoveries, findings and developments. From the appearance of early system awareness in ancient Greece, described by Aristotle in his Metaphysics, to the seventeenth century with the creation of mechanical philosophy and the discovery of classic physics and to modern system sciences and management approaches, this chapter follows the thread of an ongoing development spanning over two millennia. The historic background shows that modern complexity management does not represent revolutionary approaches for new challenges, but is based on an evolutionary process that is always driven by the needs of each specific time period.

After the reflections on the historical evolution, Chap. 5 provides a classification of complexity management by core engineering disciplines. It is often useful to transfer knowledge and methods between domains and integrate them in a new context. And while some engineering domains make extensive use of complexity management, others do not—e.g. for reasons of tradition, lack of transfer effort or differences in applied vocabulary. But when engineering domains undergo significant change, for example as we have seen in software engineering in the last decades, demand for complexity management can change too—and new methods for dealing with complexity can become important. The classification in Chap. 5 shows exemplary research work, findings and applied management approaches in identified engineering core domains, indicates their mutual overlaps in terms of similar approaches and fills in blank spots to yield a comprehensive map.

Chapter 6 builds on the knowledge mediated in the previous chapters and introduces a generic complexity management framework. This is based on structural management approaches, which have been successfully applied to complexity challenges recently. Each step of this framework is described in detail, with a specific focus on the challenge of information acquisition. The successful execution of this task is of major importance for all approaches of complexity management and presents specific hurdles. The hurdles are indicated in Chap. 6, as well as approaches to cope with them.

References

1. James, Elaine St. 1998. *Simplify Your Life*. White Plains, NY: Disney Press.
2. Schuh, Günther, and Urs Schwenk. 2001. *Produktkomplexität Managen. Strategien, Methoden, Tools*. München: Hanser Fachbuch.
3. Baldwin, C.Y., and K.B. Clark. 2000. *Design Rules—The Power of Modularity*. Vol. 1. Cambridge, MA: MIT Press.
4. Dörner, Dietrich. 1992. *Die Logik des Misslingens*. Reinbek: Rowohlt Taschenbuch.
5. Sterman, John D. 2015. Teaching Takes Off: Flight Simulators for Management Education—'The Beer Game.' http://web.mit.edu/jsterman/www/SDG/beergame.html. Accessed 29 Dec.

Introducing Complexity in Engineering

3

Complexity management requires a profound understanding of the matter of complexity. Therefore the general composition of complex systems will be introduced first, using examples to highlight important aspects and the large range of complexity. These examples are followed by a discussion of the difference between the meaning of the terms complex and complicated in the context of systems management. It will be explained why it is not just a minor linguistic difference, but a need for specific management approaches.

Despite the excessive use of the term complexity many of its definitions are either quite vague or very specific—and therefore only applicable to certain fields. For example, one mathematical definition of complexity (Kolmogorov complexity) is based on the minimal length of code required for generating a specific desired output [1]. Obviously, this is not helpful when dealing with the high-level development of an airplane or the management of a large infrastructure system. But even in engineering disciplines like systems engineering, a variety of definitions instead of a central common one can be identified [2]. Section 3.2 introduces relevant complexity definitions and indicated commonalities and differences. Special focus is placed on structural complexity, as it possesses major relevance for many applications.

Dörner mentioned that people tend to make specific errors when interacting with complex systems. For successfully managing such systems one must be aware of these mistakes, and one requires adequate methodical approaches even if the system in question seems to be non-transparent [3]. The typical problems occurring when interacting with a system are one specific consequence of complexity, which is explained in Sect. 3.3.

In the last section of Chap. 3, established engineering approaches towards complexity management are described. As each of them is covered by innumerable books, this section should only be a brief introduction, providing the basic understanding for later chapters. For example, Chap. 4 investigates the historic development of complexity management approaches. And Chap. 5 classifies the approaches, describes their differences and overlaps

© Springer-Verlag GmbH Germany 2017
M. Maurer, *Complexity Management in Engineering Design – a Primer*,
DOI 10.1007/978-3-662-53448-9_3

and links important contributors. While in Sect. 3.2 the definition of structural complexity is provided, Sect. 3.4.4 introduces dependency modeling, which gets applied in a manifold of approaches, methods and tools aiming at the management of this kind of complexity.

3.1 Composition of Complex Systems

When talking about engineering examples of complexity, often the development of the Airbus A380 gets mentioned. Especially in Europe this project has been in public awareness for a long period of time.

Without any doubt, significant technological challenges had to be met for planning and realizing the largest passenger airplane in the history of aviation. This challenge represented a new development, as it does not happen very often. Most development projects can be classified as change or adaptation developments, as they are largely based upon existing products. Here, representatively examples of new technical development challenges could be the integration of multiple new technical systems or the air conditioning for a large number of people. And as the integration of technical (sub)systems implies many interdependencies, small changes to one partial system can result in far-reaching, sometimes unpredictable and undesired consequences.

When mentioning complexity of the A380 development, first thoughts go to the technical product and its product complexity. And a technical product with such a huge scope definitely comprises much of this type of complexity. However, product complexity did not pose the only challenge in the A380 development: realizing the product required an adequate organization, e.g. with distributed development teams at several locations. The project size and the fact that it has been a new development resulted in a large organization size—and this organization formed a structure with numerous interdependencies, tremendous information flows and high dynamics.

The organization executes processes, e.g. the integration of a large number of customer requirements into the product. Since customer acquisition was of major importance for the new product, even in late stages of the development adaptations were still being conducted. And such integration of requirement-driven adaptations caused changes to the technical system and partly resulted in unforeseen impact (change propagation) because of the numerous interdependencies in the system. This impact resulted in laborious and costly rework, which also resulted in severe project delays.

Besides product-, organization- and process related complexity, another highly relevant source of complexity for the project was the embedding of the system into the environment it was designed to operate in. For example, passenger boarding processes have never before been designed for the large passenger capacity of an A380. Thus, procedures but also technical support systems (e.g. passenger bridges) had to be rethought and redesigned. And all these auxiliary processes and products have to be embedded into the airport system as a whole. A large number of interdependencies exist between the subsystems, which form the

Complex product
- Aerodynamics, position control
- Safety instructions
- Passenger comfort
- ...

Complex organization
- Distributed development team
- Several production plants
- ...

Complex embedding
- Passenger boarding
- Baggage and cargo handling
- Emergency cases
- ...

Complex process
- Integration of customer requirements at late process stages
- ...

Wiring in an Airbus A380*

Passenger bridges connected to an Airbus A380**

Fig. 3.1 Airbus A380 as a complex system, *Vitaly V. Kuzmin (http://vitalykuzmin.net/?q=node/248), CC BY-SA 3.0 (http://creativecommons.org/licenses/by-sa/3.0), via Wikimedia Commons; ** Hakilon, CC BY 3.0 (http://creativecommons.org/licenses/by/3.0), via Wikimedia Commons

greater airport system and have been impacted by changes required for embedding the A380 into it.

Figure 3.1 summarizes the mentioned aspects of complexity for the Airbus A380 project. Obviously, it would be easy to identify additional aspects of complexity; but also this exemplary description allows identifying one major characteristic of complexity: the large amount of interdependencies between system elements can produce unpredictable impact to the system (change propagation) when new system elements are implemented or existing elements or dependencies are adapted. For a person interacting with a system, complexity appears as a lack of logical match between inputs to and outputs from the system. Predictions about the system output based on input measures become hardly predictable, and control and management of such complex systems become challenging.

Another example of a complex system is illustrated in Fig. 3.2. Most people would agree that city infrastructure represents a complex system. However, the aspects that make the system complex are not always clear at first glance. In fact, infrastructure is a good example

Function of individual traffic

- Transportation of goods
- Passenger transportation
- Rescue operations

Extraordinary effects

- Spontaneous traffic jams
- Accidents and breakdowns
- Commuter/holiday traffic
- Mass events

Scope for design

- Design and layout of roads
- Signage
- Traffic light circuit

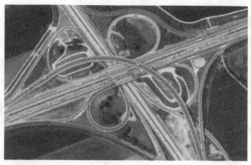

Large motorway junction*

Traffic jam on a motorway**

Fig. 3.2 Transportation as a complex system, *Bayerische Vermessungsverwaltung—http://www. geodaten.bayern.de (http://vermessung.bayern.de/open data), CC BY-SA 3.0 (http://creativecommons. org/licenses/by-sa/3.0), via Wikimedia Commons, **Alexander Blum (www.alexanderblum.de), via Wikimedia Commons

for a complex system of systems, which comprises for example energy, transportation, telecommunication and information, water and waste infrastructure systems. Those sub-systems are interconnected, and for example a failure in the energy system can lead to tremendous disruptions in the transportation and telecommunication systems.

When thinking about the complex aspects of a transportation system, often major motorway junctions with their confusing road layout come to mind. Figure 3.2 contains such a typical photograph. Even if the many interconnected roads cannot be comprehended in a first impression, this alone does not make the system complex. Roads and their junctions can be modeled as a structure of nodes and edges. And without further impact this structure is not characterized by any dynamics. Structural changes would include road construction (also inducing a low degree of dynamics) but also redirections, which can quickly change the usage of the structure. And this usage is decisive for the complexity of the transportation system. While the road structure remains mainly stable, fluctuations of traffic on individual roads induce high dynamics. Passenger traffic shows fluctuations daily, weekly as well as seasonally. And extraordinary, predictable effects like major cultural and

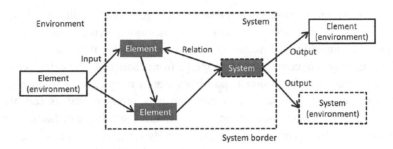

Fig. 3.3 System elements and interdependencies

sports events superimpose on those regular fluctuations. Unpredictable impact to the system occurs due to accidents, breakdowns or severe weather, which also burden the transportation system.

After introducing two real-world examples of complex systems, Fig. 3.3 depicts the basic composition of a system. It possesses a system border, separating the system itself from its environment. Inside the borders the system consists of elements, which are connected by interdependencies (relations). Section 3.4.4 will show that those elements and relations can be classified into groups by their meanings. Interdependencies also exist between elements of the system, and elements and systems in the environment outside of the system borders. The internal and external relations are decisive for the system's reaction to changes from the system's environment, either from external elements or other connected systems. In the other direction, a system produces changes that reach over the system border to impact external elements or systems.

This description of a system meets the definition of structural complexity, as introduced next in Sect. 3.2. And dynamic complexity requires system structures as a functional basis. Dynamics are not modeled in the generic system depiction of Fig. 3.3—but the interdependencies, which connect the elements, represent the paths along which dynamic effects and processes proceed. Knowing about the structure already provides information about possible dynamic behavior, e.g. because of feedback loops (which can result self-energizing effects) or bottlenecks (which can be critical because impact propagation gets channeled) [4]. System dynamics (see Sects. 3.4.3 and 4.3.5) builds upon system structures, as do many approaches and methods in systems engineering (see Sects. 3.4.2 and 4.3.4)—for example when designing system architectures or managing interfaces.

A system can also contain one or more other systems. In such a case one talks about a system of systems. In fact, both systems described above represent a system of systems. For example, the electrical system and the turbines form systems located within the entire system of the Airbus A380. Similarly, the roads and the guidance system (traffic lights, signs etc.) form their own systems within the general transportation system. For a comprehensive introduction to systems of systems, see e.g. [5].

The network built from elements and relations in Fig. 3.3 only indicates one static state of a system. As mentioned above, system dynamics models dynamic behavior based on such structures using additional modeling elements like stocks, flows and time delays. In addition, dynamic system complexity can emerge from changes to the existence of system elements and relations. In product portfolios for example, new components and possibilities of combining those components into new product variants can occur over time. And other components may disappear from the portfolio, e.g. because suppliers stopped production. In organizational structures the dynamic changes to system elements and relations can occur even more frequently and quickly than in technical systems. The collaboration and communication between employees in a company is constantly evolving, so that official organization structures often differ significantly from de facto structures. This has to be kept in mind when analyzing complex system structures.

Obviously the more relations and elements a system comprises of, the more non-transparent it becomes. Non-transparency in this context means that it is not clear how the system inputs are correlated with its outputs. The outputs cannot be predicted simply based on the inputs.

When applied to engineering tasks, the definition of the term complexity is often not very precise, compared to the way mathematical definitions are (see Sect. 3.2). In daily business, complex processes, products and organizations represent challenges in engineering. And even mixtures of these complexity domains meet, for example, in complex projects.

While most challenges in those engineering domains do not meet mathematical complexity definitions, people who have to interact with the systems experience that they do not understand the outcome based on their input to the system—and they call this as complexity. An example could be a situation where engineers are confronted with a product development process that led to extensive project time overruns when applied. As a countermeasure to those overruns, more resources could have been assigned to the process execution, expecting a reduction in process run time as consequence. If, however, the process time then would not decrease as expected (or would even increase) the engineers would experience a mismatch between system input and output, and would constitute a lack of system understanding. This exemplary development process appears to be complex to the person who is responsible for managing it. In other words, in the engineering context a system is often called complex if one cannot predict the system's output based on the given input.

The basis of complex engineering systems is the quantity of system elements and interdependencies—and on this basis, impact propagation and dynamic behavior takes place. The system elements can be classified into groups of similar objects often called system domains. Such domains can for example be process steps, product elements and organizational units. Interdependencies exist between elements within one domain as well as between elements of different domains. While identification and classification of system elements can help to improve system understanding, it is important not to reduce complex challenges to isolated system perspectives (see also Sect. 4.1.2 for the historical background on reductionism and Sect. 3.3 for typical failures when interacting with complex systems). Complexity results from the interaction in networks of elements.

In most modeling approaches, system elements receive close examination while the origin of system dependencies often gets neglected. However, many different dependency types appear in a complex system and it is important to differentiate between them for proper interpretation [4].

Distinction Between Complicated and Complex Systems

Especially in descriptions of industrial use cases, the terms complicated and complex often appear in the same context and are even used as synonyms. However, both terms represent different kinds of challenges and require specific methods for solving them. Incorrect characterization of a challenge and the subsequent mis-application of methods can be counterproductive. For this reason a more detailed consideration of complicatedness and complexity is helpful.

A typical example for a complicated challenge is searching for a needle in a haystack. This definitely represents a difficult, laborious task, but reaching the solution is only a question of effort. The more time and the more people work on the task, the higher the probability of a faster solution. The task can be parallelized so that several people search smaller haystacks. In general formulation that means that a complicated problem can be divided into smaller and less complicated problems, which can be processed independently. Maurer (according to Ehrlenspiel) mentions that the individual capability is linked to the complicatedness of a task stating that "the term complicated system describes the subjective difficulty in interaction with technical systems that often depends on one's personal abilities". Ehrlenspiel indicates that "identical situations can be complicated for one person, but not for another" [4] according to [6].

In contrast to the "needle in haystack problem", the forecasting of world climate represents a complex challenge. Such challenges are characterized by high dynamics and are difficult to subdivide into smaller, more manageable tasks. The reason therefore is that the elements of the complex system are highly interrelated. Development of water temperatures in the ocean is related to cloud formation, air movement, rainfall etc. Inadequate simplification by extracting specific aspects can easily lead to wrong system models, as important system impacts and outputs get neglected. In contrary, as high as the haystack hiding the needle may be, the system elements do not possess any relevant interdependencies—and therefore the system can be subdivided.

Thus, the large amount of elements interconnected by relations is a significant characteristic of a complex system, where the emphasis lies more on the relations than elements. It is not purely the number of product components, process steps or organizational roles that cause complex system behavior, but instead it is their degree of mutual connectivity. A dense interconnectivity of elements makes a system non-transparent to a person interacting with it and results in momentum of that system. And though as high as a hypothetical haystack may be, the system is transparent as the task, inputs and outputs are easy to understand, and system reactions to user interactions are predictable.

Complicated and complex challenges call for application of different approaches and methods. One approach towards complicated problems has already been mentioned:

increasing applied resources. As a complicated challenge can be subdivided into smaller, independent work packages, more resources can accelerate finding a solution.

Interestingly, the same approach can cause additional problems when applied to a complex challenge. In general, it is a good idea to use more resources; in the use case of the world climate, different people are assigned to model and analyze water temperatures, wind, cloud formation etc. But because all these aspects of the complex challenge are interrelated, the pieces of work are interrelated too, resulting in the new creation of organizational complexity. The engineering discipline of systems engineering tackles such challenges, where a complex technical problem has to be considered as a more holistic challenge with additional process and organizational complexity as the consequence of work distribution.

For the example of the world climate one could argue that simplified forecasting models exist. Furthermore, regional subsystems are created for forecasting on country or city basis. That would mean that reductionism can be successfully applied. In fact, reducing the modeling of a complex system goes along with the risk of neglecting relevant aspects—which then can result in wrong analysis and prognosis results. The variation between different climate predictions in practice and the deviation between assumptions of former models and experienced climate in reality gives good evidence of this effect.

The examples mentioned above show the importance of correct classification of problems as being complicated or complex, because incorrect classification can lead to wrong measures being taken. In fact, it is a common mistake to counteract the appearance of a complex problem with increased resources—without preparing to manage the increase in process and organizational complexity. Figure 3.4 summarizes the general distinctions

A pile of hay roles, now find the needle...*   Climate map of the world

Complicatedness
- Problem subjectively difficult
- Solution reachable by „hard work"
- Characteristic of system perception

Complexity
- System irreducible
- Unforeseen developments
- System characteristic

Fig. 3.4 Complex versus complicated systems, *Scott Bauer, U.S. Department of Agriculture, via Wikimedia Commons

between complicatedness and complexity. And the following sections provide definitions and layout types as well as approaches towards the management of complexity.

3.2 Complexity Definitions

Even though the term "complex" has become common word for describing many situations, its correct meaning is not easy to describe due to many aspects and different perspectives. However, if complexity shall be managed it is important to understand its origins, relevance and impacts.

The term complex originates from the Latin word stems com which means together, and plectere which means weave. Thus the combination complex can be translated as interwoven. Ashby states that "[. . .] there are complex systems that just do not allow the varying of only one factor at a time—they are so dynamic and interconnected that the alteration of one factor immediately acts as cause to evoke alterations in others, perhaps in a great many others" [7]. This description of complexity already contains the most relevant aspect: because of interconnections between elements in a dynamically behaving system, simple one-to-one effect chains rarely exist. And so changes to one element can result in avalanche-like impacts.

Scientific fields dealing with complexity have different perspectives, which results in a variety of definitions of complexity. Even within the engineering field not all aspects of complexity are commonly shared and standardized definitions that serve all fields of application do not exist [8].

The perhaps most precise definitions can be found in mathematics and computer science, because they permit the deduction of exact complexity measures. With these measures two systems can be directly compared in terms of their complexity and thresholds that trigger specific actions can be set. The required computing time or the minimal size of software code (Kolmogorov complexity) can be used as complexity measure, if the problem can be mathematically formulated [9]. Unfortunately, this kind of complexity definition and measure is not applicable to systems, which cannot be fully modeled by mathematical means. For typical complex engineering systems this is the case.

Mainzer categorizes different types of complexity from a mathematical and computer science perspective [1]. He aggregates the determination of complexity by code size or computation time as computable complexity. Besides this, notions of information complexity and dynamic complexity exist (see Fig. 3.5).

Information complexity (entropy) comprises the phenomena of noise, describing the fact that parameters oscillate with no clear timely behavior. Depending on the frequency, those oscillating effects are named white, pink, red and gray noise. Noise effects are often illustrated by the sound that results from electrical oscillations in an amplifier. However, the same effect occurs in many other situations and systems, e.g. with stock prizes or car traffic

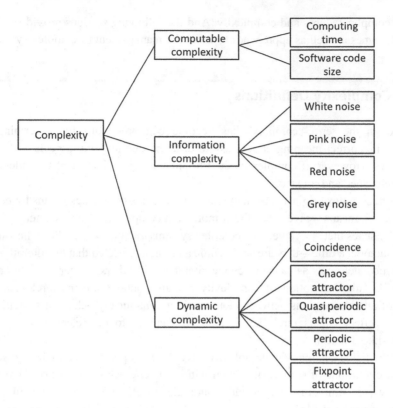

Fig. 3.5 Classification of complexity from a mathematical/computer science perspective [1]

on highways [10]. Mainzer explains: "1/f b spectra [describing the cure of a periodic signal] represent the pattern for distinguishing the different forms of signal noise in the world. [...] Signals of time series also provide hints towards self-organizing complex structures [...] [and] secular trends [...]. Time series analyses with 1/f b spectra are independent from specific systems and [...] can be applied to all kinds of dynamic systems" [1]. A well-grounded introduction to noise effects and their relevance for complexity is given by [10].

Degrees of dynamic complexity can be identified depending on the attractor, which is applicable for a complex system. An attractor is a state a dynamic system gets "attracted to" in the long run. A fixed-point attractor represents a state of equilibrium that remains unchanged. Non-linear complex systems can also reach periodically changing equilibrium states (periodic and quasi-periodic attractors) as well as turbulent or even random states (chaos attractors, coincidence) [1].

After introducing his classification of complexity types, Mainzer highlights that this should not be seen as an approach of reductionism (see also Sect. 3.2). "The structures of complex systems cannot be reduced to their single elements, but can only be explained by their collective interaction" [1]. This picks up Aristotle's famous statement that "whole is greater than the part" (Aristotle, cf. Euclid, Elements, Book I, Common notion 5).

Herbert A. Simon, a Nobel Prize laureate in 1978, defines a complex system as "one made up of a large number of parts that interact in a nonsimple way. In such systems, the whole is more than the sum of the parts, not in an ultimate, metaphysical sense, but in the important pragmatic sense that, given the properties of the parts and the laws of their interaction, it is not a trivial matter to infer the properties of the whole" [11]. Simon describes the significance of a hierarchical system structure for complex systems, saying that a hierarchic system is "a system that is composed of interrelated subsystems, each of the latter being, in turn, hierarchic in structure until we reach some lowest level of elementary subsystem".

Simon's parable of the two watchmakers became a famous and often cited exemplification for the benefits of hierarchical structures for the evolution of complex systems. In this parable, both watchmakers build complex mechanical watches composing of 1000 parts each. One watchmaker (Hora) architects his watches based on 111 subassemblies on three levels, with each subassembly consisting of ten parts. On the lowest level the 1000 basic parts are arranged in 100 assemblies. Those 100 assemblies are further aggregated into ten higher-level assemblies, which then form the entire watch. The other watchmaker (Tempus) builds the whole product as one single assembly of 1000 parts. Now it is assumed that every time a watchmaker has to interrupt his work (e.g. for taking a phone call) the currently unfinished assembly falls apart and has to be reassembled. With a simple quantitative analysis, Simon shows that Tempus loses much more work when being interrupted and that for him the probability for successfully finishing the assembly of a watch is ridiculously low compared to Hora [11].

Simon transfers the findings from his parable into the evolution of complex systems stating that "the time required for the evolution of a complex form from simple elements depends critically on the numbers and distribution of potential intermediate stable forms". So, "complex systems will evolve from simple systems much more rapidly if there are stable intermediate forms than if there are not. The resulting complex form in the former case will be hierarchic" [11].

Concerning the dynamics of complex systems, Simon explains that "hierarchies have a property, near-decomposability, that greatly simplifies their behavior". Near-decomposability means that "interactions among the subsystems are weak, but not negligible" and "intra-component linkages are generally stronger than inter-component linkages. This fact has the effect of separating the high-frequency dynamics of a hierarchy—involving the internal structure of the components—from the low-frequency dynamics—involving interaction among components" [11].

Additionally, in terms of comprehending complex systems, Simon states that "empirically, a large proportion of the complex systems we observe in nature exhibit hierarchic structure." Furthermore, "if there are important systems in the world that are complex without being hierarchic, they may to a considerable extent escape our observation and our understanding" [11].

A basic classification of engineering complexity is shown in Fig. 3.6. Market complexity can be seen as a major source of complex challenges, because market conditions and adaptations can hardly be influenced by enterprises. Market complexity can result for

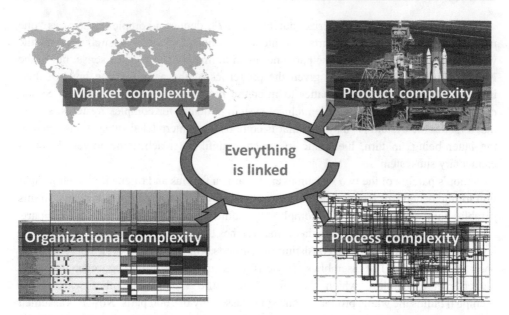

Fig. 3.6 Complexity fields in engineering, NASA/Frank Michaux, via Wikimedia Commons

example from a large variety of customer requirements that have to be fulfilled. In addition, laws, regulations or regional and linguistic peculiarities can create boundary conditions that add to the market complexity. From a company's point of view market complexity is also called external complexity, because of its origin outside of the company's direct influence.

The external complexity (as seen from the company's perspective) is faced by an internal complexity that results from the company's product portfolio. This includes combination possibilities among the variety of components, which lead to product specifications that shall fulfill the external complexity. Implementation of modular concepts, building block design, platforms and interface design represent examples for challenges in the field of product complexity.

The complexity of products and product portfolios of companies is often directly linked to the existence of process complexity. For example, the increase of product functions and components can create the need for more development process steps. And those steps are interrelated and need to be coordinated. Consequently, the company's process flowchart can become more complex due to increasing product complexity. Further constraints like decreasing development time, international product portfolios or distributed development approaches can increase process complexity even more.

Organizational complexity is also interlinked with the complexity types mentioned above. Managing complex products and product portfolios and executing complex processes requires adequate organizations. "Conway's law" [12] described this fact almost 50 years ago: He stated that when organizations design systems, those designs are similar to the organization's communication structures. This statement links product structures with organizational structures.

With the knowledge about Conway's law it is interesting to investigate the appearance of structures in organizations and products. Whereas functional and matrix-oriented structures have become popular over the last decades, hierarchical structures are still common in many organizations. On the other side, many products became highly networked structures that do not follow a hierarchical approach. Therefore organizational design is an ongoing challenge in modern product development.

It is important to consider the interrelation between the four aforementioned complexity classifications for determining the kind of a specific complex challenge. The origin of complexity and its appearance or perception are not necessarily determined in the same field. For example, an enterprise can possess a significant amount of complexity based on its comprehensive product portfolio. However, the originating cause of this complexity can sometimes be found in the markets the enterprise delivers to. It is important to identify the origin of the complexity in order to assess its value. That means a complex product portfolio does not possess any value in itself, but it can be the reason for high amounts of effort for the enterprise. In the context of an existent market complexity however, product complexity can be valuable as the product portfolio allows the delivery of the right products to this broad and diversified market.

In all four fields described above complexity results when a large number of system elements are mutually interlinked. That means it is not the pure number of product components, process steps or organizational roles, but instead their mutual dependencies that decide their complex behavior. Such systems become non-transparent for people interacting with them. A good example for a non-complex system with a huge amount of elements could be the database storing an enterprise's customers: even if many addresses might be included, this system only contains a few and obvious interdependencies (e.g. ZIP codes and cities). And if one address gets updated, deleted or added this does not have any impact to other system elements. In contrast to that, the database storing the requirements for a technical product of the same enterprise can be complex, as requirements possess many mutual interdependencies. Consequently, changes to one requirement can result in tremendous impact to many others.

In the field of product design and development, a complexity definition according to cybernetics is helpful [13]. In cybernetics simple, complicated and complex problems are distinguished. In contrast to simple problems, complicated problems are characterized by many highly interconnected parameters. Complex problems, in addition, possess high dynamics within the system. Ashby highlights that analysis methods designed for dealing with simple systems do not work for complex systems [7]. This can also be said for applying methods to complicated and complex systems, as it has been illustrated in the previous Sect. 3.1. Obviously, rules for classifying systems as simple, complicated or complex are not as explicit as pure mathematical complexity definitions. But for systems which cannot be algorithmically modeled, this represents a useful initial guideline.

Musès also categorizes complexity into three groups and from a practical perspective. Complexity I is inherited and exists in almost every system. Consequentially, it is difficult or even impossible to avoid it. Complexity II is caused by wrong handling and the usage of

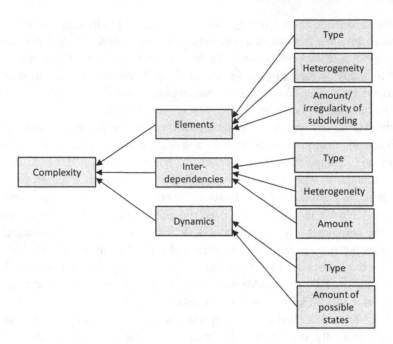

Fig. 3.7 Parameters of complexity from a product development perspective

incorrect approaches and therefore can be avoided, when better suited methods and procedures can be found and successfully applied. In contrast to that, Complexity III cannot be addressed with presently existing solutions. This type of complexity requires the creation of new and innovative methods to become manageable [14].

Lindemann defines complexity from a product development point of view [15]. He mentions the relevant parameters of complex systems to be the number of elements, their interconnections and resulting interfaces. In addition, he mentions that associated processes (e.g. design, production or distribution processes) contribute significantly to the resulting degree of complexity. Finally, Lindemann highlights that it is often important to link and integrate stakeholders into the system, which means to consider sociotechnical and not only technical system complexity. This reflects the initial thoughts of cybernetics' development, when Wiener saw the necessity to integrate operators of airplanes and air raid defense into the system modeling. In summary, Lindemann declares complexity to be dependent on the elements (type and heterogeneity, amount and irregularity of subdivision), the interdependencies (type, heterogeneity and amount) and the dynamics (type and space of possible states) [15]. This classification is shown in Fig. 3.7.

Lindemann mentions systems engineering as an approach towards complexity management [15]. A basic definition of complexity in this field has been presented by Sheard and Mostashari and is depicted in Fig. 3.8. Aspects like the size of a system, connectivity and architecture are similar to Lindemann's definition of complexity; additionally, in the systems engineering definition the aspect of environmental complexity is modeled

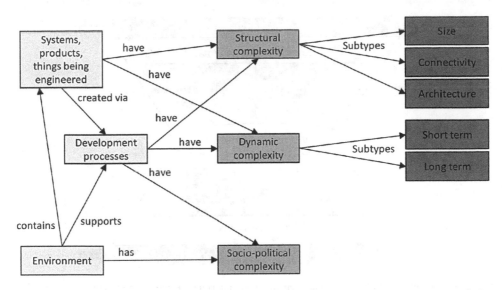

Fig. 3.8 Classification of complexity from a systems engineering perspective (adapted from [16])

explicitly. Three basic complexity types are defined—structural, dynamic and social-political complexity. While technical systems can have structural and dynamic complexity, socio-political complexity results from development processes and the environment.

In many engineering applications it can be useful to not only classify complexity by its type but by its origin. From an enterprise's point of view the separation of internal and external complexity can help in solving relevant challenges. Schuh and Schwenk introduced this perspective, depicted in Fig. 3.9 [18]. Here, internal complexity is understood similarly to Lindemann's perspective. This internal complexity emerges from the number of elements, interconnections and resulting interfaces. In addition, internal complexity can comprise process and organizational complexity resulting from an enterprise's effort of developing, maintaining and offering products or product portfolios. Thus, internal complexity results from a company's market offer.

External complexity emerges from the market requirements, e.g. the number and combination of functions requested by customers. This market-induced complexity can hardly be influenced by an enterprise and therefore represents an external source of complexity. Variant management is the challenge of matching the complexity of external market requirements with the complexity of the internal product offer. While the external complexity should be as large as possible (which means to fulfill a large variety of customer requirements), this needs to be realized with as little internal complexity as possible (which means to keep the internal efforts low) [18].

External complexity in the scenario of variant management contains a specific characteristic worth mentioning: this is one type of complexity that shall be increased, while in many other cases the objective is to decrease complexity. This external complexity can be

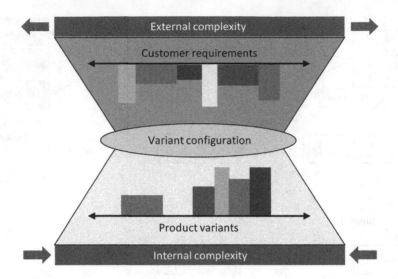

Fig. 3.9 The challenge of variant management at the interface of internal and external complexity (adapted from [17])

characterized as useful, compared to other useless types of complexity. The classification of useful and useless complexity is explained in detail in the context of a complexity management framework in Sect. 6.2.

3.3   Impact of Complexity

One basic part of a complex system is the large number of elements (variables) which are interconnected. While large is a vague term it is not possible to provide an exact number of elements that make a system complex. For example, scientists working with "systems of systems" like a city's infrastructure system would put the threshold of element numbers very high. But also systems with smaller numbers of elements can be in line with definitions of complexity.

Sheard summarizes complexity as follows: "Complexity is the inability to predict the behavior of a system due to large numbers of constituent parts within the system and dense relationships among them" [19]. That means that it is not a distinct threshold of system elements and interconnections that makes a system complex, but is instead the impact to people that results in a lack of understanding.

An impressive example is given by Browning in the context of Design Structure Matrices. These matrices represent matrix-based notations of elements and their interdependencies in a compact format. Browning mentions that even a number as low as ten elements can be difficult to oversee and manage [20]. The reason is that interdependencies between even a few elements can create high numbers of paths and loops in the system. These can cause unexpected impact (side effects), because dependency chains become long, mutually overlap or build feedback loops. Long dependency chains are hard to identify, and overlapping dependency chains and feedback loops can aggregate to intensify impact or extinguish an effect. Especially feedback loops can cause unstable system behavior and the resulting effects can hardly be evaluated without high computational effort. System dynamics is an approach specifically dealing with feedback loops, which will be introduced in Sect. 3.4.3 in the context of engineering application and in Sect. 4.3.5 from a historic perspective.

Dynamics is another important characteristic of complex systems, which is mentioned in all different definitions of the term. However, the precise specification given from mathematics (see Sect. 3.2 and Fig. 3.5) is not helpful for application to engineering challenges, which cannot be fully algorithmically modeled. Nevertheless, the impact of dynamics to complex systems can be described. Norbert Wiener, the pioneer of cybernetics, was the first to model human operators and technical devices in an integrated model, which was helpful for solving the associated challenge; but such systems turned out to be complex control problems to solve. The development of cybernetics is described in Sect. 4.3.2.

As long as no interaction happens either between the environment and the system or between elements within the system, even a high interconnectivity between multitudes of system elements does not result in effects of complexity. Interaction with the system means that information is transferred via an interconnection and this action can initiate further interactions along connected elements and dependencies. If, however, no interaction happens, the interdependencies are inactive and therefore irrelevant. In other words, effects of complexity are associated with the application of a system.

An example can highlight the significance of this statement: Most people would agree that today's smartphones represent complex systems. People think so, as they might think about the many (interconnected) electronic components or the many software applications. But if one were to use a smartphone to participate in a mobile phone throwing competition (that really exists: http://www.mobilephonethrowing.fi/), no complexity is associated with the phone. In fact, it would make no difference if one uses the phone or a brick (of same size and weight) for the competition. The absence of complexity in this (rather unusual) use case results from the fact that the application does not trigger any informational impact to the technical system. And consequentially the system elements and interdependencies are irrelevant for this case. If, however, a developer has to apply a technical update to a smartphone, the effects of complexity can easily occur. The technical measure causes impact to the system, which can spread via interdependencies to many other parts of the system. In the worst case unpredicted effects can occur.

It needs to be mentioned that complex systems do not need external input for a dynamic behavior to initiate. Dynamics can emerge in the system itself and either the specification of elements/variables can change or their interdependencies. A typical example for an internal source is a failure of single element, which can result in tremendous consequences. These consequences mean that the system produces visible output that passes the system boundary. Thus, the system is open. Closed systems do not interact with the environment; by definition they are isolated from it and every impact remains within the system. Bertalanffy classified systems according to their interaction with the environment [21]. His General Systems Theory is more closely explained in Sect. 4.2.

The large number of elements and interdependencies make a complex system non-transparent and incomprehensible to people, who can develop a fear of interacting with such a system [3, 22]. Obviously, this can significantly impact decision processes and lead to failures when dealing with such systems [3]. Dörner mentions four causes for failures when interacting with complex systems: Slowness of thinking, protection of one's own competence, minimal recording of information and fixation of attention to the actual problem only. The resulting failures can, for example, be observed and experienced in business games like the "Beer game" (see Sect. 4.3.5)—which illustrates immediate action, ineligible system or process reduction/abstraction or neglecting side effects resulting from insufficient problem understanding. Other failures have a psychological basis, e.g. endless planning without acting, solving known problems or ad hoc reactions.

Considering the possible impacts of complexity like the inability to make decisions or wrong decision-making, this points out its tremendous relevance. While uncontrolled complexity poses high risks to organizations, societies and enterprises, successful management of complexity also implies significant opportunities. For an enterprise context, Maurer mentions a lack of decision-making ability, frequent development crises and product changes as well as long development process duration as consequences resulting from complexity [23]. In addition, he describes that the effective managing of complexity can provide beneficial opportunities like increased competitiveness, successful control of large variant and product spectra and possibilities of increased product customization. And as managing complex systems is more challenging than dealing with simple systems, successfully managed complex systems imply a significant hurdle for copycat products and competitors entering the market.

Vester describes the risks of complexity for human societies. He states that due to an increasingly complex world (mentioning unemployment, dramatic environmental changes, stock market crashes and military conflicts), even well-planned interventions can lead to fatal consequences because of feedback loops and time delays [24].

Complexity is an integral part of many systems and seems to be required for realizing higher states of development. Complexity can be found in biological systems, and impressive functionalities such as that which is delivered by the human brain seem to be not achievable with simple system design [22, 25]. Thus, complexity can be naturally required for a system to work, and an indiscriminate strategy of complexity avoidance can be harmful. This does not imply that all complexity is necessary and helpful. In fact, it is important to distinguish useful and useless complexity and treat both kinds accordingly. This will be further explained in Sect. 6.2.

In 2012 the study "Mastering Complexity" from Camelot Management Consults tackled the relevance of complexity for the economy [26]. Eighty-three percent of the top managers surveyed (more than 150 participated in the study) mentioned that the degree of complexity in their enterprises was too high. Eighty-nine percent said that complexity increased within the last 3 years and 76% said they expect complexity to further increase in the near future. The study mentions that enterprises could raise their EBIT by 3–5% if complexity management is successfully implemented. While the Camelot survey included top managers from a variety of fields, the industrial sector alone has seen an increasing relevance of complexity and its management for the future. Fifty-nine percent of the managers expect increasing sales figures for offers of complete system solutions.

3.4   Established Complexity Management in Engineering

Offering a large product portfolio to the market often implies complexity. Therefore, it is interesting to have a closer look at one of the most famous companies with an almost inconceivably large spectrum of products—Amazon.com. The company started out as an online book store, and then became the largest retailer in the USA. If one only considers the number of products, the ability to economically offer such a large product portfolio would seem to be highly complex.

While the success of Amazon is highly impressive, the product portfolio (considered as a system) does not possess many interdependencies. Changes to one product do not influence other products, the business processes or the organization. And new product requests from customers (the market) can be served with additional product offers that do not impact the enterprise's processes or organization.

The company Spreadshirt became popular by their offer of customized shirts with short delivery time. This business of minimal order size can be realized, because product development (creating the customized shirt) is not burdened by significant interdependencies. The same business model with for example customized combustion engines would most likely fail, as each customization would impact many other parts of the product as well as system elements in the process (production, testing) and organizational field. The challenge of managing system interdependencies explains why product customization approaches so far focus on simple systems only.

Mass customization, as a blend of the terms mass production and customization, has obtained much interest over the last 20 years. However, offered products like NikeID (www.nike.com/NIKEiD) or Reebok Custom (www.reebok.com/us/customize) only allow selecting from a predefined set of color options or decorations. Those customization tools represent configuration approaches and do not fulfill the mass customization idea of "producing goods and services to meet individual customers' needs with near mass production efficiency" [27]. Product customization approaches are applied on complicated but not complex systems.

For tackling complex problems in the engineering context, four main approaches have become established. These techniques are partly based on the same groundwork and mutual influence. The historic context will be explained in Chap. 4, indicating the steps from system awareness to modern complexity management. In the following sections these approaches shall be briefly introduced in terms of objective, functional scope and application context.

One fundamental approach of complexity management is not introduced in a separate section, as it comprises part of most other techniques—systematic visualization. System elements and their interdependencies contribute significantly to a system's complexity. If those system structures can be externalized, users can more easily understand and simulate system behavior. In engineering many visualization techniques get applied, and depending on the specific field some specifications have become quasi-standards, e.g. event-driven process chains or SysML models [28]. However, Sect. 3.4.4 introduces dependency modeling, which is based in large part on visualization techniques.

3.4.1 Operations Research

Operations research is often misunderstood as a collection of mathematical methods. While it is the application of many of these methods, it cannot be reduced to a purely computational approach. Quantitative models and methods are used to help find an optimal decision for complex challenges. Human decisions and non-quantifiable arguments are not considered in this part of the approach, but best integrated in the subsequent analytical part of the approach. Thus, operations research is not a straightforward solution-finding process, but provides support for optimal decision-finding.

Operations research is characterized by the cooperation of disciplines like mathematics, economics and computer science, and can be subdivided into sub-branches. Linear optimization, transport optimization, combined optimization and dynamic optimization are a few prominent ones. Also game theory often gets applied in operations research (see Sect. 4.3.6) [29, 30]. In general, deterministic and stochastic approaches are distinguished in operations research.

The main objective of operations research is to describe decision problems by optimization and simulation models, and to develop an algorithm that can be applied for solving the problem. Problems are classified by criteria like degree of information, (types of) target functions, and constraints and solvability. Therefore, decision, optimization and simulation models get applied [30]. Operations research focuses more on the system states than on the system structure and often applies experiments.

Based on the definition and objective of operations research, six phases of application exist. These phases do not represent isolated, straightforward process steps, but are highly interlocked with each other or can be worked on in parallel if required.

– Formulate the problem
– Develop a mathematical model
– Deduce a solution based on the model
– Validate the model and solution
– Supervise and adapt the solution
– Implement the solution

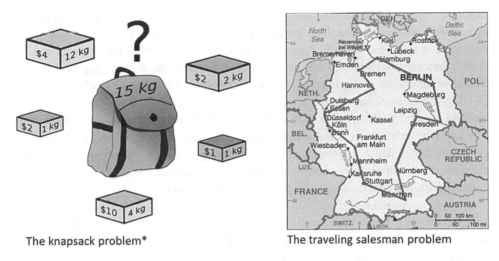

The knapsack problem*   The traveling salesman problem

Fig. 3.10 Knapsack problem and traveling salesman problem, two complex challenges in operations research, *Dake, CC BY 2.5 (http://creativecommons.org/licenses/by/2.5), via Wikimedia Commons

A popular example for the application of operations research on a complex challenge is the knapsack problem [31]. The knapsack problem is based on the theoretical use case that one has to fill a backpack with objects of maximum value. The available objects have different sizes and values while the backpack provides a limited amount of storage space. While the example is hypothetical, this challenge appears in a large variety of real-world problems. For example loading trucks with payloads or organizing the workload of computing centers can be modeled by this abstract problem. The knapsack problem represents a challenge of computational complexity and can be mathematically described and solved—e.g. by dynamic programming approaches.

Another popular problem that can be solved by operations research methods is the traveling salesman problem. Here when given a set of predefined places, the target is to find the shortest path visiting each place exactly once and returning back to the starting point. Despite the fact that the situation and the constraints can be clearly described mathematically, it still represents a complex computational problem, as defined from a mathematical perspective (computable complexity, see Fig. 3.5). The problem can be solved using heuristic and approximation algorithms [32]. Figure 3.10 (right side) shows the shortest path to visit the 15 largest cities of Germany in a closed loop. This path represents one out of 43,589,145,600 possibilities.

3.4.2 Systems Engineering

Products and modern production facilities are often highly complex, large-scale projects. They contain an unmanageable combination of technical systems and impact factors: mechanics, electrics, energy, control systems, hardware and software, humans and machines, logistics and communication, customers and suppliers. More and more product requirements are accompanied by increasingly faster development processes and strict constraints concerning budget, quality and time to market. With these conditions in mind, it becomes obvious that the development of for example a passenger airplane, an innovative luxury car or any other large system represents a huge task, which can only be solved by interdisciplinary teams. The required knowledge for such projects is extremely manifold, multidisciplinary and extensive. While it is impossible to concentrate this knowledge in only one or a few people, there is the necessity of project planning and decision-making, which requires well-informed people with a comprehensive project overview. The need for such system specialists increased with the beginning of the age of industrialization and the associated technological advances.

The systems engineering approach was developed to meet these upcoming needs for system specialists. Initial applications were conducted after the Second World War. The most famous use case became the application of systems engineering for Project Apollo, the US aerospace program in the early 1970s.

The original term systems engineering can be traced back to Bell Laboratories in the United States in 1940, where it became mentioned in the context of weapons system development. Related and partly integrated approaches and terms are e.g. systems architecture, systems engineering management and systems design. Those terms are partly used as synonyms, and make a unified definition of systems engineering rather challenging. The common denominator of all these approaches is that for solving a complex problem, the problem gets disassembled into smaller parts and reintegrated later into a final solution [33].

In German-speaking countries different translations of systems engineering are in use, which can be misleading. While the English term systems engineering is more commonly used (e.g. the German Chapter of INCOSE (International Council of Systems Engineering) is named "Gesellschaft für Systems Engineering") [34], terms like Systemtechnik (German for system technology) and Technische Systemanalyse (German for technical system analysis) are also applied. While Technische Systemanalyse focuses on technical system design, Systemtechnik typically includes associated procedures like project management. Then terms like system management or system control are also applied [35]. Saynisch classifies systems into three categories: material or object systems, process-related systems and target systems. He sees systems engineering as planning and designing technical (object) systems and representing the developmental and creative part, which is complemented by project management as the controlling part [35]. This differentiation between systems engineering and project management represents a modern view, while older definitions saw project management being a building block within the systems engineering approach, e.g. [36].

Systems engineering can be applied to a technical product or a superior man-machine-system, including the application of a product. According to Bluma, systems engineering is the integration of different components into a technical system. The components are considered by their function for the entire system [37]. Bluma mentions that the system is not constrained to the technical components only, but can also integrate non-technical elements of system organization and system use into the analysis. This allows treating each engineering problem with a systems engineering approach. This approach integrates widely fragmented sub-disciplines into one engineering approach and formed the concept of a systems engineer. Several methods and models of systems engineering originate from cybernetics, e.g. black-box principles, block diagrams and statistical system analyses [37]. The holistic approach of systems engineering is well-suited for being applied in solving complex problems. Systems engineering tools make use of qualitative as well as quantitative methods.

3.4.3 System Dynamics

Many questions of society, corporations and organization are complex challenges, as the system behavior is often nonlinear—cause and effects cannot be fully understood and system impacts can be unpredictable. For such challenges often the approach of system dynamics is applied. System dynamics is a mathematical approach that models complex systems using flows between system elements that can form feedback loops and stocks [38]. Developed in the 1950s by Jay Forrester, it became a powerful computer-supported modeling approach that is widely used for complex problems. The most famous application is the modeling of exponential growth for the world, described in the publication The Limits to Growth [39].

In system dynamics, four basic concepts can be differentiated: The core is one concept which describes information feedback, meaning that interactions between system components can be even more important than the components themselves. The system behavior is mainly resulting from feedback loops between its elements. Feedback loops can be classified into positive and negative feedback. They possess a specific structure and impact can occur with delay [40–42]. Automated decision-making is the second concept, which is based on a military approach of strategic, long-term decisions including their mathematical modeling [42]. Computer simulations enable a basic understanding of complex system behavior and represent the third concept of system dynamics. For example, different management approaches or market expectations can be assumed and their potential consequences can be estimated by computer simulations [38, 42]. The fourth concept then is the use of digital computers, which is required for conducting simulations and representations of complex systems [40].

An important part of system dynamics is the modeling of a complex, dynamic system. Forrester proposed a six-stage modeling process, guiding the user from the problem to a solution. Forrester, as well as his former student Barry Richmond, were both focused on quantitative modeling approaches, as they thought it was the only way to understand dynamic system behavior. Geoff Coyle was another student of Forrester and later founded the system dynamics group at the University of Bradford. Coyle had a contrary opinion concerning modeling approaches, which led to a long-lasting scientific dispute [43, 44].

The main aspects and differences between quantitative and qualitative approaches in system dynamics modeling shall be briefly explained. First of all, the type of modeling depends on the selected representation form, which can be distinguished into four categories: verbal descriptions, cause-and-effect diagrams, flow diagrams and mathematical equations [45].

According to Ossimitz, verbal descriptions of system models are easy to understand but rather unprecise. Consequentially, these descriptions can only usefully be applied to qualitative system model descriptions [45].

Cause-and-effect diagrams are typically composed of graphs built up from nodes and edges. Here, system elements are represented by nodes, while the impact from one system element to another gets modeled by edges connecting two nodes. Edges can be specified by plus or minus symbols to provide a more detailed description of the cause-and-effect relationship between the two nodes. A plus symbol indicates a monotonically increasing relationship, while a minus symbol indicates a monotonically decreasing relationship. If all edges that form a closed feedback loop are specified by plus or minus symbols, then a restraining or escalating behavior can be identified. Ossimitz explains that it is not required to specify system elements in a cause-and-effects diagram with numbers. However, some system elements can already imply a quantitative specification. Ossimitz mentions the example of car traffic, where either cars on the streets in general or a specific amount can be expressed [45].

The third form of representation is the flow diagram. This can be seen as an extension of the cause-and-effect diagram by different types of system elements and relations. System elements are separated into state/level/stock variables, flux/rate/flow variables and auxiliary variables. Standardized symbols are applied for representing system elements in flow diagrams [44, 46]. The meaning of this differentiation is that timed processes often require distinguishing between variables that possess a value at a certain point of time and variables that possess a value after a specific time. One example could be the balance sheet of an enterprise, which comprises state variables for a specifically selected day. In contrast to that, the profit and loss statement comprises flow variables related to the flow rate of 1 year. For reasons of easier simulation, continuous processes are also modeled with state and flow variables just like in the discrete case.

The fourth option of representation in system dynamics is the application of mathematical equations. These equations only comprise quantifiable variables and typically build the basis of a numerical simulation [45].

In system dynamics, qualitative as well as quantitative models can be applied. In practical applications, often the problem situation gets acquired and structured first by applying qualitative modeling. Quantification represents a follow-up step, which allows for simulation-ready models that can be used for scenario analyses [47]. Also Jay Forrester holds the opinion that qualitative approaches should be supplemented by acquiring a subsequent quantitative system dynamics model. He argued that qualitative methods do not meet the requirements of dynamic simulations [43, 44].

3.4.4 Dependency Modeling

Dependency modeling does not represent a separate discipline, approach or technique. Here the term is used as an aggregation for those methods aiming to support the management of structural complexity. When these methods were developed, visualization of system structures was one of the main objectives. With the increase in computational power and the big data approaches, visualization became more abstract and computational methods replaced manual system interaction. Nevertheless, structural modeling and its associated methods of complexity management are still widely in use.

Already with the rise of cybernetics, graphs were applied for indicating system dependencies. However, matrix representations were more common because they were already integrated in applied mathematics. Graph visualizations can be very intuitive, and many dependency models make use of them. Especially process models often get visualized with graphs, and some became quasi-standards for whole industry branches, for example event-driven process chains. Browning and Ramasesh provide a comprehensive overview of applied process models in product development [48]. Graph depictions of dependency networks typically require much drawing space and can become more confusing the large the networks become. This is mainly because of crossing and long-range dependencies, which make it difficult to understand and interpret the network.

Matrix models of dependency networks make it easier to apply computational methods, which is one of the reasons for their intensive use. In addition, matrices provide the basis for systematic dependency acquisition processes. Alexander was the first to document the acquisition of dependencies in a matrix with a symmetric arrangement of elements on both axes, which created a pattern for depicting interdependencies in the matrix cells [49]. For this specific matrix form, techniques of reordering elements and dependencies have been developed, which allows one to analyze and optimize dependency networks. These square matrices and the methods of their manipulation get aggregated under the term Design Structure Matrix (DSM) and have been developed since the early 1960s [50]. Methods have been further developed since then and many industry applications are documented [51].

One of the main applications for DSM is the identification of useful modules in a system by rearranging matrix columns and rows, see e.g. [23, 52]. As well, process or task sequences can be streamlined using the same technique. Yassine and Braha mention, "The DSM approach to managing complex development projects is an information exchange model which allows the project or engineering manager to represent important task relationships in order to determine a sensible sequence for the tasks being modeled" [53]. A comprehensive introduction to DSM methods is provided by Eppinger and Browning as well as Lindemann et al. [8, 51].

Figure 3.11 shows a typical DSM with elements listed in the row and column headings, and interdependencies between the elements indicated by dots in the inner matrix cells. The location of dependencies shows potential for module-building, as most dependencies are

		1	2	3	4	5	6	7	8	9	10	11	12	13	14	15	16
Radiator	1		X														
Engine fan	2	X				X											
Heater core	3																X
Heater hoses	4																
Condenser	5		X				X		X								
Compressor	6					X			X	X							
Evaporator case	7																X
Evaporator core	8						X	X		X							X
Accumulator	9						X		X								
Refrigeration controls	10																
Air controls	11																
Sensors	12																
Command distribution	13																
Actuators	14																
Blower controller	15																X
Blower motor	16			X				X	X							X	

Fig. 3.11 Design Structure Matrix (DSM), adapted from [54]

clustered in a block. However, in terms of finding an optimal module structure the elements need to be further relocated in the matrix order.

DSMs represent a highly compact form of dependency modeling. They are definitely not as intuitive as a graph representation; however more elements can be depicted with DSMs. And once familiarized with the methods, one can visually identify specific structural patterns, e.g. clusters, straightforward dependency chains, feedback loops or isolated system areas.

The DSM is complemented by the Dependency Mapping Matrix (DMM) representing a matrix with two different types of elements on either axis [55]. This format allows representing dependencies between two types of elements, while the DSM documented dependencies of elements within one type. The format of a DMM is very common in many other applications and also outside of the field of engineering. Examples of applied names are "cause and effect matrix" or "interface structure matrix" [56, 57]. However, the DMM approach also comprises analysis and optimization methods for system dependency networks. For example the clustering of a structure represented in a DMM allows for simplifying the dependency management between two types of elements in a system [58–60].

Figure 3.12 on the left side shows the modeling capabilities of a DSM in graph format, which is limited to one element type and one dependency type only. The DMM enlarges the possibilities of system modeling to two types of elements, while still only one dependency type can be depicted (illustrated on the right side of Fig. 3.12).

Looking at Fig. 3.12 it becomes obvious that DSM and DMM can only model small system parts, e.g. the internal component structure or associations of people to tasks or

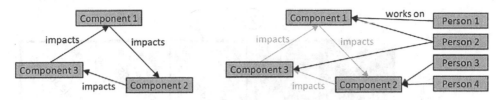

Fig. 3.12 DSM and DMM representation capabilities in graph format

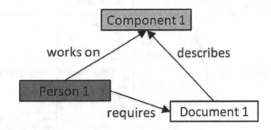

Fig. 3.13 Several element types and dependency types form the reality in many systems

components. However, system complexity is very much about the holistic perspective including various element types. One of the central claims of cybernetics was the interdisciplinary approach, which naturally includes many element and dependency types. Figure 3.13 illustrates a very simple system composed of three elements belonging to three different element types. And between them three dependencies of different meanings exist. Such a system could not be represented nor analyzed with the means of DSM or DMM. These approaches only extract single system views for further investigations, e.g. the component network.

A more holistic approach on dependency depiction is provided by the Multiple-Domain Matrix (MDM), which allows modeling several element and dependency types and deriving specific system views for closer analysis and optimization [4]. An abstract illustration of an MDM is shown in Fig 3.14. This matrix shows the element types (domains) and not the single elements. Each of the five domains (components, people, data, processes and milestones) contains a distinct number of elements. The inner fields of the matrix in Fig. 3.14 represent sub-matrices containing the dependencies between the elements of the types listed in the column and row headings. The words in the abstract matrix fields indicate the meaning of the dependencies.

The upper left matrix field (indicated by 1) contains change dependencies between components. Thus, this matrix represents a DSM, as do the four other matrices along the diagonal of the MDM. The next matrix to the right (indicated by 2) contains dependencies between components and people; here a dependency means that a specific component is "processed by" a specific person. Because of the two different element types on the two axes, the matrix is a DMM, and so are all the off-diagonal matrices in an MDM.

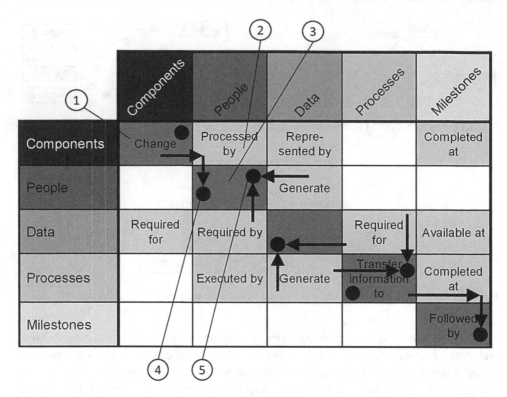

Fig. 3.14 Multiple Domain Matrix (MDM)—holistic system structure analysis

The matrix field indicated by 3 is a DSM field for the element type "people". The fact that no dependency meaning is noted in the field shows that no dependencies have been acquired and so far the matrix field is empty. However, dependencies between people can be computed based on other dependencies in the MDM. As examples, these possibilities are indicated by the black arrows leading into the matrix field. Such matrix computations result in indirect dependencies, which aggregate dependency information from two (or even more) element types into one system view.

The first computation uses the DSM of component dependencies and the DMM between components and people, and is indicated by 4. The resulting (indirect) dependencies between people indicate that two specific people are connected because they work on components that are linked by a change impact. If the modeled people represent product developers working on parts of a technical system, this system perspective shows their mutual dependencies based on their component responsibility. Interpreting this network could be used, for example, to optimize the organization of group meetings. Figure 3.15 shows an exemplary network of people that can result from the matrix computation.

Another computation for creating a people DSM is indicated with the second chain of arrows and the indicator 5. In this case, the DSM is computed from the information "people who generate data" and "data required by people". While both networks are DMMs, they

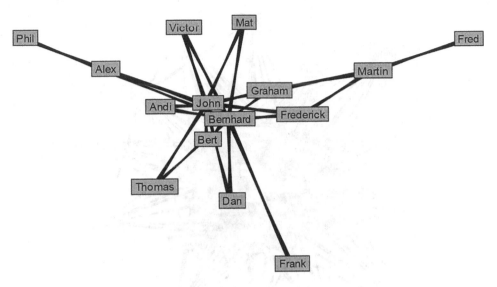

Fig. 3.15 Designers' dependencies based on their responsibility for components [23]

have different meanings and consequentially different dependency networks. An example result of the people network is shown in Fig. 3.16. The meaning of this computed network is that people are interconnected because a person generates data that is required by another person. This network showing the provision and request of data between e.g. product developers could be applied in improving the team organization in case of staff fluctuation. The computed dependency network provides a significantly different perspective, with a different group of people in the core of the network than in the network before. This indicates the importance of defining and creating the correct system views for answering specific questions.

The short example shows the major benefits of an MDM, which are capturing different types of system elements in a comprehensive model as well as computing and selecting specific system perspectives. These perspectives represent DSMs or DMMs, which means that the well-established methods of analysis, optimization and visualization can be applied. Maurer provides a comprehensive introduction to MDMs [60]. Industrial use cases are described by Lindemann et al. as well as Eppinger and Browning [8, 51].

Dependency modeling represents an important step in many approaches towards complexity management. Information visualization might be one of the popular fields of application, as many impressive network depictions have been created. The website "visualcomplexity" shows hundreds of large-scale networks with many different graph visualization approaches [61]. For most of these projects, information acquisition has been conducted by automatic tracking or data mining. Thus, the effort for collecting and preparing a large amount of system elements and dependencies was manageable. If information has to be acquired from people by interview—e.g. for documenting the

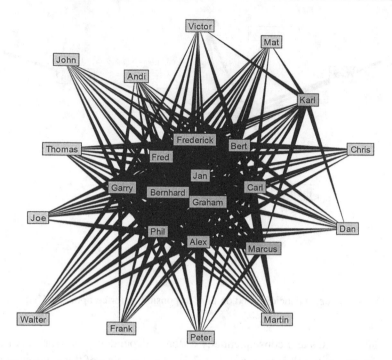

Fig. 3.16 Designers' dependencies based on their request and provision of data [23]

development processes, tasks and exchanged data between them—then dependency modeling also provides the approach of systematic data acquisition. More details on information acquisition can be seen in Sect. 6.4.

One characteristic of many complex challenges in engineering is the lack of transparency. This is especially because dependencies are often unclear or even unknown. And people involved in a project might have different understandings of the dependencies without knowing about the discrepancies. Dependency modeling makes system dependencies explicit and therefore stimulates discussions and a common understanding of situations and systems.

References

1. Mainzer, Klaus. 2008. *Komplexität*. Paderborn: Wilhelm Fink.
2. INCOSE. 2009. Systems Engineering Complexity Types.
3. Dörner, Dietrich. 1992. *Die Logik des Misslingens*. Reinbek: Rowohlt Taschenbuch.
4. Maurer, Maik S. 2007. *Strcutural Awareness in Complex Product Design*. Munich: Dr. Hut. http://nbn-resolving.de/urn/resolver.pl?urn:nbn:de:bvb:91-diss-20070618-622288-1-1.
5. Maier, Mark W. 1998. Architecting Principles for Systems-of-Systems. *Systems Engineering* 1 (4): 267–284. doi:10.1002/(SICI)1520-6858(1998)1:4<267::AID-SYS3>3.0.CO;2-D.

6. Ehrlenspiel, Klaus. 2007. *Integrierte Produktentwicklung—Methoden Für Prozessorganisation, Produkterstellung Und Konstruktion.* 3rd ed. München: Hanser.
7. Ashby, Ross. 1961. *An Introduction to Cybernetics.* Fourth imp. London: Chapman & Hall.
8. Lindemann, Udo, Maik Maurer, and Thomas Braun. 2009. *Structural Complexity Management— An Approach for the Field of Product Design.* Berlin: Springer http://medcontent.metapress.com/ index/A65RM03P4874243N.pdf.
9. Kolmogorow, A.N. 1965. Three Approaches to the Quantitative Definition of Information. *Problems Information Transmission* 1(1): 1–7.
10. Peak, David, and Michael Frame. 2013. Komplexität—Das Gezähmte Chaos. Springer.
11. Simon, Herbert. 1962. The Architecture of Complexity. *Proceedings of the American Philosophical Society* 106(6): 467–482 .http://nicoz.net/images/ArchitectureOfComplexity.HSimon1962.pdf
12. Conway, M. 1968. How Do Committees Invent? *Datamation* 14: 28–31.
13. Wiener, Norbert. 1948. *Cybernetics or Control and Communication in the Animal and the Machine.* New York: Technology Press.
14. Musès, C. 2002. Simplifying Complexity: The Greatest Present Challenge to Management and Government. *Kybernetes* 31: 962–988. doi:10.1108/03684920210436282.
15. Lindemann, Udo. 2005. *Methodische Entwicklung Technischer Produkte.* Berlin: Springer.
16. Sheard, Sarah A, and Ali Mostashari. 2010. A Complexity Typology for Systems Engineering.
17. Schwanitz, D. 1999. *Bildung. Alles, Was Man Wisen Muss.* Frankfurt: Eichborn.
18. Schuh, Günther, and Urs Schwenk. 2001. *Produktkomplexität Managen. Strategien, Methoden, Tools.* München: Hanser Fachbuch.
19. Sheard, Sarah. 2012. Assessing the Impact of Complexity Attributes on System Development Project Outcomes.
20. Browning, Tyson R. 2001. Applying the Design Structure Matrix to System Decomposition and Integration Problems: A Review and New Directions. *IEEE Transactions on Engineering Management* 48(3): 292–306.
21. Bertalanffy, Ludwig v. 1973. *General System Theory: Foundations, Development, Applications.* 4th ed. New York: George Braziller.
22. Pruckner, M. 2002. 90 Jahre Heinz von Förster. Die Praktische Bedeutung Seiner Wichtigsten Arbeiten. Malik Management Zentrum St. Gallen.
23. Maurer, Maik S. 2012. Komplexitätsmanagement Für Die Industrielle Praxis—Komplexe Systeme Und Ihre Eigenschaften. Technical University of Munich.
24. Vester, Frederic. 2007. *The Art of Interconnected Thinking—Ideas and Tools for Tackling Complexity.* München: Mcb Verlag.
25. Holland, John H. 2000. *Emergence—From Chaos to Order.* Oxford: Oxford University Press.
26. Camelot Management Consults. 2012. Mastering Complexity.
27. Kaplan, Andreas, and Michael Haenlein. 2006. Toward a Parsimonious Definition of Traditional and Electronic Mass Customization. *Journal of Product Innovation Management* 23(2): 168–182. doi:10.1111/j.1540-5885.2006.00190.x.
28. Delligatti, Lenny. 2013. *SysML Distilled: A Brief Guide to the Systems Modeling Language.* Boston, MA: Addison-Wesley Professional.
29. Beckmann, M., H. Gehring, K.-P. Kistner, C. Schneeweiß, G. Schwödiauer, H.-J. Zimmermann, and T. Gal, eds. 1992. Grundlagen des Operations Research 3: Spieltheorie, Dynamische Optimierung Lagerhaltung, Warteschlangentheorie Simulation, Unscharfe Entscheidungen. 3. Band. Springer.
30. Hillier, J.H., and G.J. Lieberman. 2004. *Introduction to Operations Research.* 8th ed. Boston, MA: McGraw-Hill Higher Education.
31. Kellerer, Hans, Ulrich Pferschy, and David Pisinger. 2004. *Knapsack Problems.* Berlin: Springer.

32. Applegate, D., R. Bixby, V. Chvátal, and W. Cook. 2006. *The Traveling Salesman Problem.* Princeton, NJ: Princeton University Press.
33. Weigel, Annalisa. 2000. An Overview of the Systems Engineering Knowledge Domain. MIT. http://web.mit.edu/esd.83/www/notebook/sysengkd.pdf.
34. Gesellschaft Für Systems Engineering E. V. 2015. www.gfse.de. Accessed 29 Dec.
35. Saynisch, Manfred. 2008. Systems Engineering Und Projektmanagement in Deutschland. Projektmanagement Aktuell 3: 1–9. www.pmaktuell.org/uploads/PMAktuell-200803/PM_3_08-025lang.pdf.
36. Daenzer, W.F., and F. Huber. 1999. *Systems Engineering: Methodik Und Praxis.* 10th ed. Zürich: Verl. Industrielle Organisation.
37. Bluma, L. 2004. *Norbert Wiener Und Die Entstehung Der Kybernetik Im Zweiten Weltkrieg.* Münster: LIT.
38. Forrester, Jay W. 1963. *Industrial Dynamics.* Cambridge, MA: MIT Press.
39. Meadows, D.H. 1972. *The Limits to Growth.* Reissue. Verlag Signet.
40. Hahn, F. 2006. *Von Unsinn Bis Untergang: Rezeption des Club of Rome und Grenzen des Wachstums in der Bundesrepublik der frühen 1970er Jahre.* Freiburg: Albert-Ludwigs-Universität.
41. Richardson, G.P. 1999. *Feedback Thought in Social Science and Systems Theory.* Waltham, MA: Pegasus Communications.
42. Tohum, M.H. 2008. System Dynamics—Betriebswirtschaftliche Anwendungsgebiete. Universität Koblenz-Landau.
43. Coyle, R.G. 2000. Quantitative and Qualitative Modeling in System Dynamics: Some Research Questions. *System Dynamic Review* 16(3): 225–244.
44. Kapmeier, F. 1999. Vom Systemischen Denken Zur Methode System Dynamics. Universität Stuttgart.
45. Ossimitz, G. 1991. Darstellungsformen in Der Systemdynamik. In *Anschauliche Und Experimentelle Mathematik*, ed. H. Kautschitsch, 175–184. Wien: HPT.
46. Coyle, R.G. 1996. *System Dynamics Modelling—A Practical Approach.* London: Chapman & Hall.
47. Gabler Wirtschaftslexikon. 2015. http://wirtschaftslexikon.gabler.de/Archiv/143837/system-dynamics-v5.html. Accessed 29 Dec.
48. Browning, Tyson R., and Ranga V. Ramasesh. 2007. A Survey of Activity Network-Based Process Models for Managing Product Development Projects. *Production and Operations Management* 16(2): 217–240. doi:10.1111/j.1937-5956.2007.tb00177.x.
49. Alexander, C. 1964. *Notes on the Synthesis of Form.* Cambridge: Harvard University Press.
50. Steward, Donald. 1981. The Design Structure System: A Method for Managing the Design of Complex Systems. *IEEE Transaction on Engineering Management* 28(3): 79–83.
51. Eppinger, Steven D., and Tyson R. Browning. 2012. *Design Structure Matrix Methods and Applications.* Cambridge, MA: MIT Press.
52. Fernandez, C.I.G. 1998. *Integration Analysis of Product Architecture to Support Effective Team Co-Location.* Cambridge, MA: Massachusetts Institute of Technology.
53. Yassine, Ali, and Dan Braha. 2003. Complex Concurrent Engineering and the Design Structure Matrix Method. *Concurrent Engineering* 11(3): 165–176. doi:10.1177/106329303034503.
54. Pimmler, Thomas U., and Steven D. Eppinger. 1994. Integration Analysis of Product Decompositions, no. September.
55. Danilovic, Mike, and Tyson Browning. 2004. A Formal Approach for Domain Mapping Matrices (DMM) to Complement Design Structure Matrices (DSM). In Proceedings of the 6th Design Structure Matrix (DSM) International Workshop. Cambridge, UK: University of Cambridge, Engineering Design Centre.

56. Allen, T.T. 2006. *Introduction to Engineering Statistics and Six Sigma—Statistical Quality Control and Design of Experiments and Systems*. London: Springer.

57. Kusiak, A., C.-Y. Tang, and Z. Song. 2006. Identification of Modules with an Interface Structure Matrix. ISL_04/2006. Iowa.

58. Danilovic, Mike, and H. Börjesson. 2001. Managing the Multiproject Environment. In Proceedings of the 3rd Dependence Structure Matrix (DSM) International Workshop. Cambridge: Massachusetts Institute of Technology.

59. ———. 2001. Participatory Dependence Structure Matrix Approach. In Proceedings of the 3rd Dependence Structure Matrix (DSM) International Workshop. Cambridge: Massachusetts Institute of Technology.

60. Maurer, Maik S. 2007. *Structural Awareness in Complex Product Design. Lehrstuhl Für Produktentwicklung*. München: Dr. Hut.

61. Lima, Manuel. 2011. *Visual Complexity—Mapping Patterns of Information*. New York: Princeton Architectural Press.

History of Complexity Management

4

Globalization, interconnectivity and advancing technology are some of the reasons often mentioned when investigating the origin of increasing complexity. It might look like the challenge of complexity and the solutions supplied by complexity management arose from questions and constraints in modern times. However, there are circumstances, trends and developments that led to the modern form of complexity management. A historical classification shall provide a better understanding and explain the fundamentals of modern complexity management. Over a long period of time, many scientists have worked on the phenomena of complexity. And due to approaching challenges, developed approaches, methods and tools have been applied to important challenges in practice.

When looking for significant increases of complexity in history, one can reach back to several important events, which planted the seeds for further societal, political and techno-logical development. For example, Schwanitz explains the end of the 7 Years' War in 1763 as the beginning of complexity [1]. While the start of free worldwide trade is obviously relevant in terms of upcoming complexity, developments during the Second World War and the second industrial revolution maintain perhaps even more importance for today's state of complexity management. While the Second World War accelerated many inventions and (product) developments, the second industrial revolution generated the need for effective methods for controlling and managing large projects and systems.

Scientific and methodical knowledge about modern complexity management has been aggregated over a time span of approximately 70 years. With increasing complexity and new challenges this knowledge has been constantly developed. Today, the problem is often not a lack of methodical approaches but missing knowledge about implementation details when dealing with complex challenges in the industrial context.

Thinking about the phenomenon of complexity and dealing with this challenge can be traced back to ancient Greece. Beginning with Plato and especially Aristotle, complexity became an important topic of interest. From the seventeenth century on, exceptional

© Springer-Verlag GmbH Germany 2017   43
M. Maurer, *Complexity Management in Engineering Design – a Primer*,
DOI 10.1007/978-3-662-53448-9_4

mathematicians like Isaac Newton and in the nineteenth century philosophers like Emanuel Kant were dealing with complexity and changed the view of the world. In various disciplines they thought about newly upcoming challenges and developed knowledge and approaches to tackle the challenges.

If one name needs to be associated with modern complexity management, it is Ludwig von Bertalanffy, who developed major parts of system theory. His work still represents fundamentals for most work in the field of complex systems. Based on Bertalanffy's understanding of systems and their characteristics, a variety of approaches for dealing with complexity have been developed—from which complexity management emerged as its own discipline.

Generally, the history of science is in close relation to contemporary, social, political and military challenges. This also accounts for complexity management. A major part of basic science and method development for complexity management have been created in the context of the Second World War. Norbert Wiener created the fundamentals of complexity science—named cybernetics [2]. After that pioneering but still abstract theoretical work, applicable methods and instruments for solving complex challenges were created: operations research, system dynamics, systems engineering and game theory. These approaches emerged from different questions and therefore tackle different problem areas. And for example, the more holistic approach of systems engineering integrates the possibilities of more focused approaches than operations research or game theory. Significant challenges like those presented by the Cold War and the beginning of astronautics acted as catalysts for those developments on complexity management. The innovations resulting from solving these challenges have in common that they are trans- and interdisciplinary science approaches and that they transcended system borders that existed in the centuries before.

Cybernetics meant a revolution in thinking and paved the way for today's "cyber" world. In several aspects, modern data processing and the computer are based on cybernetics. And those developments led to increasing system dependencies and more dynamics in products, processes and organizations—which meant an increase of complexity. On the other hand, cybernetics also provided the basis for solving modern complexity challenges.

Operations research, system dynamics, systems engineering and game theory have been further developed for the application of complexity management in fields like economics. And as mentioned before, they are partly interconnected. Game theory represents an instrument used within operations research, which has been developed at the time of the Second World War for optimizing the British radar monitoring. Since then it has been adopted to a multitude of problems, many of them in economics. Operations research is a quantitative method, i.e. it requires an algorithmic problem description and aims at optimizing specific target parameters.

System dynamics is a cycle-based methodology for the simulation and analysis of non-linear behavior in complex systems. Therefore, a system structure is created and extended by stocks, flows and feedback loops. System dynamics is often applied to management tasks and economic challenges, but has successfully been transferred to many other systems as well, using qualitative as well as quantitative models.

Systems Engineering provides a comprehensive framework for designing complex technical systems. It includes many methodologies along the whole system development process and life cycle, e.g. requirements engineering, quality and risk management or system modeling. Systems engineering overlaps with many related approaches, e.g. software engineering (also software systems engineering) or project management. At first, systems engineering came up in the 1940s in applications of the Bell Telephone Laboratories and became a significant attraction when successfully applied to the Apollo Program and the Space Shuttle program in US aeronautics.

It is difficult—and sometimes even impossible—to clearly distinguish the developments of complexity management methods in their historical context, especially because of the interdisciplinary approach of these methods. Already the founders of the fundamental ideas were working in interdisciplinary groups and transferred concepts from one field to another. In fact, systems thinking and complexity management generally require holistic approaches, as reducing problems to specific, isolated aspects would neglect the essence of complexity.

Pioneers in system thinking and cybernetics like Ross Ashby and Heinz von Foerster disagreed with the fragmented scientific world, which was the status quo before their time of research. Mostly because of historical reasons, scientific disciplines were clearly separated at this time. But system thinkers did not see these separations as being helpful for describing complex phenomena. Ross Ashby was a psychiatrist and discussed questions on cybernetics with psychologists, physiologists, mathematicians and engineers in the Ratio Club [3]. Ludwig von Bertalanffy, the originator of the General System Theory was a biologist and Norbert Wiener, who first introduced cybernetics, was a mathematician and philosopher.

These initial thoughts indicate that complexity management is not an invention made at the end of the twentieth century. It is based on long-term developments originating from different disciplines. The following sections will give a deeper insight into the historical development of system thinking, which is a fundamental basis for dealing with complexity. With the background of system thinking, the next sections describe the historical development of complexity management approaches.

During the Second World War scientists worked on possibilities for controlling complex systems—based on scientific thinking that had been developed for more than 2000 years. In the 1940s, Norbert Wiener worked on solutions for controlling complex systems. He introduced the term cybernetics in his book titled Cybernetics: Or Control and Communication in the Animal and the Machine [2]. The term cybernetics is derived from the ancient Greek word for steerman (kybernétes). Here, the link to the ancient Greeks can be seen as a reference to the basis of scientific thinking, as they have been constituted by Socrates, Plato and especially Aristotle. According to Laszlo, increasing complexity led to system science and to cybernetics as approaches toward controlling complexity [4].

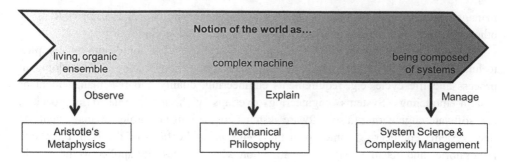

Fig. 4.1 From system awareness to complexity management

Figure 4.1 aggregates the main perspectives of system interaction into a logical sequence. In the beginning, living, organic ensembles but also societies became recognized as systems. And their observation resulted in complex questions of life and the early system thinking, as introduced by Aristotle. Progress in disciplines like mathematics, physics and astronomy and the possibilities to precisely describe new findings by laws were the fundaments of the mechanical philosophy. The means of mathematical and physical descriptions for simple technical constructs were applied to explain the perceived world and the composition and behavior of their beings as complex machines. The subsequent notion of the world as being composed of systems and the increasing need to manage these systems prepared the groundwork for system science and finally complexity management.

4.1   The Emergence of System Thinking

Inspired by Socrates (469–399 BC) and Plato (428/427–348/347 BC), Aristotle is considered as the founder of science [1]. He made the famous statements that "[...] the totality is not as it were, a mere heap, but the whole is something besides the parts..." (Aristotle, Book VIII, 1045a. 8–10) and "[...] the whole is greater than the part" (Aristotle, cf. Euclid, Elements, Book I, Common notion 5). This thought can be seen as the basic idea of system thinking, which requires awareness of not only the parts, but also their interdependencies [5]. The statement also includes the awareness that important aspects of a system get lost when subdividing it and considering only its components. Aristotle applied a general definition of a system, which has a purpose and an objective. The system-based worldview of Aristotle has largely influenced later scientific development until the Renaissance, when it was challenged for example by the mechanical philosophy.

Starting in the sixteenth and seventeenth centuries many discoveries have been made, which questioned the worldview that before had been taken for granted. This was the starting point for the mechanical philosophy. The core of this scientific revolution was driven by people like Copernicus, Galilei, Descartes, Newton and Kepler, and comprised the analysis of elements by disassembling materials into smaller pieces and by giving them mathematical descriptions. The fundamental thought of this concept was that the behavior of a system can be fully understood if the characteristics of its parts are known and understood [6]. Descartes thought of the cosmos as a giant machine following eternal laws. Newton formulated this worldview with mathematical descriptions. In order to belong to natural science he demanded things to investigable by experiments, exactly measurable and describable by mathematics. Newton developed a theory of mechanics that was formulated with mathematical exactness, which allows one to derive particular cases that can be empirically validated [7].

One striking example of the thinking in the mechanical philosophy is the "Digesting Duck", created by Jacques de Vaucanson in 1739. Vaucanson was an engineer who invented and built innovative automata, e.g. for looming. The "Digesting Duck" was an impressive mechanical construction, which exemplifies the belief in mechanical explications for nature and the world in general. Figure 4.2 shows an illustration of the Digesting Duck clearly accentuating the mechanical thinking at that time. While this illustration is widely known and nicely highlights the mechanistic concepts, it must be mentioned that Vaucanson's real machine worked differently and was unfortunately destroyed in the late nineteenth century.

In the eighteenth century, Romanticism raised opposition against the mechanical Cartesian views. For example Goethe and von Humboldt saw in nature a pattern of interrelations within a sorted system. This approach gets close to modern system thinking [8]. Goethe questioned Newton's hypothesis that things only exist when they can be described in mathematical form. Goethe emphasized that it is not only about the analytical description, but also about its composition—and that the existence of things is not directly bundled to the possibility of quantification [8].

Emanuel Kant contributed important thoughts to the later emerging worldview of the Romantic period. He distinguished between self-reproducing and self-organizing

Fig. 4.2 Illustration of the "Digesting Duck"

organisms and machines. While the components of machines do only contribute to an entire function, components of an organism do also produce each other mutually—they exist because of their common existence [8].

In the following sections the scientific thinking about systems will be introduced. Starting with the ancient Greeks and the fundamental observation of systems, the application of mathematical descriptions to the explication of nature as reductionistic thinking to contradictory approaches of modern system thinking will be detailed.

4.1.1 The World View of Aristotle

The ancient Greek philosophers have laid out the basis for our scientific thinking in the fields of mathematics, nature, society and politics. Socrates, Plato and Aristotle stand out for having significantly influenced scientific thinking to the modern day. Socrates was the teacher of Plato, and Plato was the teacher of Aristotle.

Socrates was the first to notice that religion was no longer an adequate instrument for leading a state, which becomes more and more interlinked. These considerations and the new thinking were continued by Plato. He described the Ouroboros as a self-sustaining being without outward relations—which already shows similarity to modern system definitions [9]. In continuation of this thinking, Plato's student Aristotle developed metaphysics and teleology and today is referred to as the founder of science [1].

Aristotle (384–321 BC) conducted empirical studies and intensely studied living processes and organisms. He saw the universe as an organic, living and spiritual entity. He considered form and matter as being connected and that they can only be separated by means of abstraction. Aristotle no longer thought of an entity as only being composed of its parts. This awareness is documented in the famous statement that the whole is greater than the sum of its parts. And it can be considered as the basic idea of a system [5]; it gains many of its characteristics from interactions between its ingredients. A living organism is more than just an aggregation of its parts; it represents an assembled entity with greater functions like self-preservation, which cannot be association with specific elements of the entity. One can say that science started in the Antique when the great thinkers discovered fundamental principles of systems.

Aristotle's Metaphysics (a collection of some of his scientific papers) mentions aspects of organic systems, which cannot be originate from their parts only. Aristotle discovered that creatures act purposefully by targeting specific objectives. Many modern scientists picked up Aristotle's fundamental ideas in the concept of holism (from Greek holos "all, whole, entire") as the opposite to reductionism, which postulates that a system can be fully described by its parts [10].

Aristotle generalized his observations and postulations to a more abstract level and implemented them into his political models. In doing so he found similarities between the objectives in nature and in humans acting in society. Also in this use case the components do not sufficiently describe the entity of society, and only a holistic system view covers the objective. Thus, Aristotle was the first to create a general system concept that was applicable to different areas of life.

Martial, social and political events were often interwoven and represented complex challenges for the great ancient thinkers. More than 2000 years later, in 1946, the participants of the Macy Conferences referred to this epoch by selecting an antique word for their new science for controlling complex systems: Cybernetics.

4.1.2 Mechanical Philosophy

Aristotle's worldview dominantly influenced science for a long period of time. It took from his productive period in the fourth century BC up to the fifteenth and sixteenth centuries until his notion of the world was challenged by a significantly different approach, the mechanical philosophy. Nevertheless, it has to be mentioned that not only the reference of the term cybernetics remained as reminiscence to Aristotle's work. Many great scientists, such as Ludwig von Bertalanffy, directly picked up Aristotle's ideas for their scientific approaches in modern times.

Beginning in the late sixteenth century many groundbreaking works in mathematics, physics and astronomy were made. They challenged the view on the world as a living and spiritual universe, and raised the hope that the entire world can be explained by the new rules and analysis capabilities, which worked for many observed physical phenomena at that time. This was the beginning of the mechanical philosophy. People like Copernicus, Galilei, Descartes, Newton and Kepler advanced a scientific revolution, which focused on the analysis of elements and decomposing matter in smaller and smaller parts [5].

René Descartes (1596–1650) and Isaac Newton (1643–1727) have particular relevance for the development of the mechanical philosophy [6]. Descartes, a French philosopher and mathematician, understood the universe as a machine operating by a set of rules. According to the mechanical philosophy, the world was considered to be inanimate and living organisms to be machines. Several concepts emerged in the same time frame and are strongly related to this philosophy. Rationalism, determinism, reductionism and atomism have to be named in this context. Rationalism described the philosophical concept that only rational thinking should be applied for achieving and evaluating knowledge [11]. Determinism postulates that all events are clearly determined by preconditions—thus, knowing all influences to a situation "determines" the possible outcome [12]. Reductionism can be defined as the opposite of the holistic approach, as it is based on Aristotle's approach. Reductionism assumes that a system can be fully determined by an accurate description of its included parts. Descartes also thought of animals as being explicable by this reductionistic approach as complex machines. This leads to the notion of atomism, which represents the basic idea that everything is formed from smallest indivisible parts. This explication of observable things and effects can be traced back to an origin in the thinking of the ancient Greek philosophers [13]. All these ideas aggregated in the mechanical philosophy contributed to a different view of the world and provided explications for new findings made in science. With the new possibilities of mathematical descriptions, the mechanical philosophy advanced the view that knowing the characteristics of all parts allows understanding the whole thing [8].

Isaac Newton picked up Descartes' mechanical thinking and enhanced it by mathematical descriptions. He created a complete and mathematically formulated theory of physics, which permitted one to logically derive use cases, which then can be evaluated empirically [7]. This depicts the basic ideas of determinism, which—as introduced above—states that an exact description of the current state by physical laws allows determining each future state of the system. This way of thinking proved to be valid for a long time, as it seemed to be in accordance with findings in classical physics and astronomy.

Similar to the strict application of laws in Newton's classical mechanics, determinism became applied to phenomena of nature and even the entire universe. Thus, nature was supposed to be explicable based on distinct laws. Many scientists adopted Newton's classical mechanics, and along with it concepts like reductionism and determinism. A common hypothesis was that only matter exists, which behaves according to physical laws. This explication was transferred to all kinds of natural phenomena.

The mechanical philosophy was also adopted for explaining biological processes [14]. Prominent biologists like Rudolf Virchow, Louis Pasteur, Claude Bernard and Jacques Loeb made groundbreaking discoveries and created the impression that functions and characteristics of living organisms can be understood by applying chemical and physical principles [8]. In such a system, a separate principle of life was not considered to exist—which represented a point of criticism for disputants of the mechanical philosophy and was counteracted in concepts like vitalism.

An example for the application of the mechanistic philosophy in other areas is the so-called Taylorism, named after the American engineer and entrepreneur F. W. Taylor. Taylorism reduced the human to a "gear wheel" in a large production machine, which could be replaced by another human if required. Detailed work descriptions and target times were applied for each working task [15].

The exact, mathematical, quantifying, isolating, causal, analytical and mechanical approaches result in reductive thinking. If something could not be measured or not be described by mathematical means then it was not considered by science. One consequence of isolating small parts from a large system resulted in a fragmentation of science into more and more specialized and disconnected areas [8]. This represented a significant hurdle for future approaches towards the management of complexity—and was explicitly addressed by the pioneers of cybernetics.

4.1.3 Opposition to the Mechanical Philosophy

The beginning of Romanticism in arts, music, philosophy and literature also meant opposition to the mechanical philosophy and started in the late eighteenth century. Scientists of the Romantic period disagreed with the reductionistic thinking in the mechanical philosophy and rather referred to Aristotle's finding that the whole is more than its parts only. Romantics saw the mechanical explications as a too rational attempt to control nature. Cunningham and Jardine state that "the romantics were certainly hostile to the mechanical natural philosophy and descriptive natural history that they inherited from the Enlightenment" [16].

Also Emanuel Kant contributed to the (later emerging) romanticism. He studied the nature of living organisms and declared that they were self-reproducing and self-organizing organisms—in contrary to machines. In a machine the parts support each other, whereas the parts of a natural system exist because of each other's existence [8].

Discoveries in biology made at the end of the nineteenth century, e.g. about the behavior of living cells, could no longer be explained by a reductionistic approach according to the mechanical philosophy. The embryologist Hans Driesch (1867–1941) conducted outstanding experiments with sea urchin embryos and failed to explain the outcome by the common thinking of his time. In his experiments, Driesch removed parts of the eggs but embryos could still develop, which meant that that any single monad of the egg could develop any part of the embryo. A reductionistic approach assumed that distinct parts assemble the entity; however this did not explain his observations [17].

Driesch's scientific experiments and that of other biologists were decisive for the foundation of a scientific vitalism, which represents a concept postulating that the behavior of living organisms cannot explained by physical and chemical laws only. Rather, a separate life principle or soul has to be included in order to explain it. The vitalism predicates a fundamental difference between organic and inorganic systems [18, 19].

Even if most modern biologists reject the concept of vitalism, it contributed to the development of system understanding beyond the restricted view of mechanical philosophy with its pure reduction to system parts and its strictly linear, deterministic effects.

In the early twentieth century, opposition to the mechanical philosophy came from two significant scientific breakthroughs. Firstly, the development of a quantum theory that led to quantum mechanics explained physical behavior on the atomic level, whereas Newton's classical mechanics failed for such use cases. That is to say, the associated Heisenberg's uncertainty principle disproved strict determinism as a valid concept and introduced a statistical understanding of causality instead of linear cause-and-effect chains.

In the context of quantum developments, Einstein introduced his general theory of relativity in 1915. This theory postulated that energy and matter can be transformed into each other and therefore do not represent two different concepts. This disproved Newton's idea of inanimate matter affected by immaterial forces.

4.2   Bertalanffy's General Systems Theory

Ludwig von Bertalanffy (Fig. 4.3) was born in 1901 in Austria and studied the history of art, philosophy and biology, and since 1934 held several professorships, i.e. in Vienna, London, Canada and the United States. Bertalanffy had significantly contributed to the idea

Fig. 4.3 Ludwig von
Bertalanffy (1901–1972).
BCSSS—Bertalanffy Center for
the Study of System Science,
Vienna, Austria, BCSSS-Archiv:
Ludwig von Bertalanffy
Teilnachlass 2 [LvB-TN-2],
Porträtfoto Ludwig von
Bertalanffy, June 1966

of a systems approach for explaining complex phenomena. And today he is seen as one of
the most important protagonists of system theory in the twentieth century. His works on the
General Systems Theory (GST) represent the basis for definitions and terms in all forms of
system science [20–22].

Bertalanffy introduced a new paradigm of science as an alternative to the mechanical
worldview, which has been adopted by many scientists of his time. Bertalanffy criticized
reductionism, deductive procedures and the isolated consideration of singular phenomena
as being insufficient for understanding real-world systems. Nevertheless, he did not reject
the mechanical philosophy completely. Formal models should still be a part of system
science, as they represented successful approaches for the depiction and explication of
isolated scientific application.

Bertalanffy wrote, "Since the fundamental character of the living thing is its organization, the customary investigation of the single parts and processes cannot provide a
complete explanation of the vital phenomena. This investigation gives us no information
about the coordination of parts and processes. Thus the chief task of biology must be to
discover the laws of biological systems at all levels of organization. We believe that the
attempts to find a foundation for theoretical biology point at a fundamental change in the
world picture, this view, considered as a method of investigation, we shall call 'organismic
biology' and, as an attempt at an explanation, 'the system theory of the organism'." [23]
according to [24]. And he added later: "Recognized 'as something new in biological
literature' [. . .], the organismic program became widely accepted. This was the germ of
what later became known as general systems theory" [24].

Bertalanffy did not consider linear, deductive descriptions as being sufficiently powerful
for modeling biological systems. He argued that in such systems no independent use cases
exist, but all aspects and phenomena are interlinked. Bertalanffy described a system instead
of single phenomena, whereas the system definition comprised a quantity of elements and
their interrelations. He further distinguished open and closed systems [21].

A closed system represents a special case of a general system and correlates with the approach of the mechanical philosophy with its descriptions of isolated (closed) phenomena. For such a closed system it is assumed that the initial state and final state are directly related and can be logically described. For example, this is the case in a chemical reaction, which starts with distinct settings and ends in a chemical equilibrium. Closed systems are independent from their environment and the interconnectivity of comprising elements can be described mathematically [21, 25].

According to Bertalanffy, open systems are characterized by constantly and non-predictable exchanges of energy and matter between the system and its environment; these exchanges pass across the system border. Open systems cannot reach a state of equilibrium with maximum entropy, as this would mean that no further exchanges of energy or matter would pass the system borders. But the internal variability enables the system to reach a dynamic equilibrium, which means the system changes its dynamics and its state without losing its general structure. In addition, in open systems the same final state can be reached starting from different initial states. Open systems can form self-organized, complex systems, which develop a specific higher-ranked structure without impact from the environment [5, 21, 25].

Bertalanffy considered the entity as being the result of continuous interaction between the system parts. And the development of life would be the result of processes like differentiation, specialization, and centralization in combination with an increase of complexity. For the creation of a commonly applicable model for complex systems, Bertalanffy took statistical thermodynamics as a guideline. Also in this discipline the differentiation between open and closed systems exists. Statistical thermodynamics, as formulated by Boltzmann, tries to describe the system without considering the impact of its elements in detail. Nevertheless, it is possible to create specific system laws [25, 26].

In his General System Theory, Bertalanffy formulated common laws in social, physical and biological systems based on a methodic holism approach. In this context he postulated that principles exist, which, if they have been discovered in one specific field, can be transferred and applied to others. Such principles are, for example: complexity, equilibrium, feedback and self-organization. Bertalanffy's General System Theory predicted that his stated system laws were applicable to scientific disciplines like biology and sociology, even if those cannot be classified under the framework of physical or chemical laws (see also [27]).

Bertalanffy distinguished four types of equilibria: dynamic equilibrium (used as an umbrella term), real equilibrium (as it appears in closed systems), steady state equilibrium (for open systems) and homeostatic equilibrium. The steady state equilibrium describes that energy and matter are exchanged over the system border but that the different streams add up to zero. The homeostatic equilibrium is an equilibrium which is reached by a secondary regulation mechanism [5, 21].

With his General Systems Theory, Bertalanffy created the necessary basis for a systems approach in complex biological systems. Wiener described this field by the name systems biology, which gained increasing importance since then [2]. Systems biologists see the organization and interdependencies as of high importance for understanding living organisms. The term system is used for organisms as well as social systems, and means an integrated entity which obtains its main characteristics from the relations between its parts. Living organisms are organized in a hierarchy, in which each subsystem forms its own entity. These subsystems can belong to a greater entity. In this context also the term organized complexity is applied [8].

4.3   Development of Complexity Management

Systems thinking has been a research topic for many scientists, who focused on explaining complex phenomena and the behavior of systems. Using the fundaments of systems thinking for the application of managing complexity was the next big step, which was approached in the first half of the twentieth century. At the time, scientists started to work on finding solutions to newly upcoming economic, social and political problems.

A small group of systems scientists were active in the time before the Second World War. However, the general political and social situation did not provide the circumstances for getting enough attention. The First World War and its negative consequences like tremendous human losses and economic crises did not prepare a fruitful ground for the development of new sciences. Finally, in 1929 the world economic crisis began and was followed by the rise of several totalitarian regimes in Europe. Especially, scientists from the German Reich often only saw the option of ending their ongoing research activities and decided to leave the country. This time meant the end of interdisciplinary and international science, and affected significantly the small group of systems scientists and their progress in those days.

4.3.1   Impact from the Second World War

In the Second World War, the multitude of new weapon systems and theaters of war made the organization of war become much more complex than ever before. The initial success of the German Wehrmacht had shown that conventional warfare did not succeed anymore. This initial course of the war made clear that new technologies became increasingly important for gaining an advantage over the opponent [28].

The changed situation of war resulted in a huge demand for new thinking and innovative technologies, which led to a scientific boom, especially in the USA and Great Britain. As the objective behind the boom was combating the enemy, science and military closely cooperated in creating new technologies, approaches and methods. Leading scientists from different disciplines started to research on effective solutions to urgent and painful challenges. The tremendous complexity of these demands showed the scientists the limits of applicability of the so far highly diversified partial sciences. They understood that the many and quickly changing problems required interdisciplinary cooperation; therefore they established collaborations, which should turn out to be fruitful.

These circumstances of war brought system sciences back in the spotlight. Based on systems thinking, new theories and methods for successfully handling complexity were created. The multilayered and heterogeneous challenges resulted in the development of three closely related disciplines for managing complex problems: operations research, game theory and cybernetics. While clearly related to each other, each single discipline forms its own field of research with specific problems to solve [29].

In the time after the Second World War, research in system science profited from the established close collaboration between the military, government and academia. This was the right time to transfer findings made during the war into civil applications, and to further develop and optimize them. The following sections will detail the historical questions and challenges, which resulted in new approaches and methodologies. Also the developed solution approaches will be discussed from a historical perspective.

4.3.2   Cybernetics

From the moment when it became introduced, cybernetics was designed as an interdisciplinary approach. It did not only apply mathematical and physical models, but aggregated opinions and findings of scientists from fields like mathematics, physics, economics, sociology, psychology and biology. Interdisciplinary collaboration became possible because cybernetics works at a high level of abstraction, which serves all the included disciplines. In the cybernetics approach a complex system gets described by its purpose and not by its components or specific functions and mechanisms. This was meant for improving system comprehension and reducing the model complexity. And it allows cybernetics to model the behavior of highly different systems [29].

Aerial warfare was a revolution in the Second World War. Airplanes were already used in the First World War, but not in a significant way. In WWII this new way of warfare dissolved the hitherto clear separation of the war front and homeland. Military planes started to carry battles far into the countries and caused fatal consequences for soldiers and the civilian population. The increasing threat due to aircraft bombs motivated Norbert Wiener to develop an air defense system. His findings from this development represented core elements of the cybernetics approach, for which he was one of the important founders [29].

4.3.2.1 Norbert Wiener

Norbert Wiener (Fig. 4.4) was born November 26th, 1894 as son of a Harvard professor for Slavic languages. Norbert Wiener was intellectually gifted and started a college career in mathematics as soon as he turned eleven. Later he also studied zoology and philosophy. Wiener visited and worked at the most famous universities of his time, e.g. Harvard,

Fig. 4.4 Norbert Wiener
(1894–1964), Konrad Jacobs
(http://owpdb.mfo.de/detail?
photo_id=4520), CC BY-SA 2.0
de (http://creativecommons.org/
licenses/by-sa/2.0/de/deed.en),
via Wikimedia Commons

Cambridge (UK) and Göttingen. In 1912, he completed his doctoral thesis at Harvard about mathematical logics.

During the First World War, Wiener taught philosophy at Harvard, worked as an engineer for General Electric and as an author for the Encyclopedia Americana. Later he worked as a ballistician for the US military in Maryland and started teaching mathematics at MIT once the First World War was over. In World War II, Wiener worked again for the military, especially in the field of communication and information technology. The heavy bombing of London by the German Air Force drew Wiener's attention to the development of anti-aircraft guns. Here, complexity arose from the fact that the defensive aircraft gun as well as the offensive aircraft were both controlled by humans. It turned out that this constellation was a highly complex problem of control theory. Wiener found that solving the problem required modeling the aircraft and the pilot as one integrated system, as well as the anti-aircraft gun and its operator. This new approach of system-building definitely blurred the boundaries between human and machine [2, 29]. Up to this time biological and technical systems were considered separately or, according to the mechanical philosophy, both were considered as technical systems.

Modeling a man-machine system required the cooperation of many different partial approaches. Natural sciences, engineering science as well as human behavior science had to be combined. This integration required the interconnection of these different disciplines by a discipline-spanning framework that can be commonly applied. Cybernetics was this integrated approach, which was mainly initiated by Norbert Wiener as a result from the research findings he made in different projects until the end of World War II [30].

4.3.2.2 The Scientific Approach of Cybernetics

During World War II, Wiener created the fundamentals of his new interdisciplinary and system-based science. In 1948 he published these findings in the book titled Cybernetics or Control and Communication in the Animal and the Machine [2]. Cybernetics had the main focus on controlling complex systems—and Wiener's book gained major importance for the future of complexity management [2, 31].

Besides the intensive interdisciplinary cooperation of sciences, Wiener's cybernetics also integrates technological progress into systems thinking. Wiener discovered powerful up-to-date possibilities for modeling interconnection and interaction for the man-machine. For example, his system descriptions included wave filters, calculators, automated control of assembly lines, chemical productions or even fibrillation of the organic heart [29].

Wiener's possibilities of system description enabled him to predict the impact and consequences of technical achievements to social life. At the time he introduced cybernetics, an equal treatment of human and machine in systems thinking represented a provocation, which was fought by many other scientists. However, in the 1950s technologization of common life became omnipresent. And Wiener's approach of blurring the boundaries between the human and machine became strongly supported in science [29].

In January 1950, Time magazine published the title story "The Thinking Machine", and put an illustration of the computer Harvard Mark III on the cover with the figure caption saying: "Mark III. Can man build a superman?" [32]. The illustration shows a man-machine computer in a military look (this was referring to the fact that cybernetics came from a military application field), feedback through self-control of outputs (the computer in the illustration visually "inspects" its output) and biomechanical design (a human eye and two arms connected to a computer rack). This publication was the final proof for how significant the impact from the new science became for common social life.

After World War II cybernetics got enduring attention from leading scientists. A milestone in the further development of this new cybernetics discipline were the proceedings from the interdisciplinary Macy Conferences, titled by the bio-physician Heinz von Foerster with "Cybernetics" (more about these important conferences below). This resulted in an even faster spreading in the science world.

In natural sciences cybernetics was the basis for a specific communication theory and informatics, which appears in cooperation with engineering sciences e.g. in fields like robotics, automation and artificial intelligence [30, 33]. Today, also the scientific fields of neurophysiology and genetics would be at an extremely different status quo without the preliminary work in cybernetics.

4.3.2.3 The Macy Conferences

From 1946 to 1953 the Josiah Macy, Jr. Foundation held ten interdisciplinary conferences, which are often called the "Macy Conferences". In fact, this term would officially include a much larger set of conferences over a broader period of time; however, people mostly refer only to the ten conferences on cybernetics. This specific set of conferences was an experiment on interdisciplinary research with the first conference entitled "Feedback Mechanisms and Circular Causal Systems in Biological and Social Systems" [34]. Conference titles changed over time and Heinz von Foerster (Fig. 4.5) proposed the name cybernetics. He applied it for the proceedings of the later conferences [35–38]. It is

Fig. 4.5 Heinz von Foerster
(1911–2002), U of I publicity
department, CC BY-SA 4.0-3.0-
2.5-2.0-1.0 (http://
creativecommons.org/licenses/
by-sa/4.0-3.0-2.5-2.0-1.0)], via
Wikimedia Commons

worth mentioning that the first five conferences were not systematically documented and results can only be retrieved from a few sources [34, 39, 40].

The central objective of the Macy conferences on cybernetics was to create the fundament "for a general science of the workings of the human mind" [34]. Therefore, the participants also applied the preliminary work of Norbert Wiener, who was one of the conference contributors. Besides him, a so-called "core group" of scientists attended the conferences, including e.g. the biophysicist Heinz von Foerster and the mathematician John von Neumann. Each conference was joined by additional, invited guest scientists (but also the core group changed over time), forming a highly interdisciplinary group of outstanding scientists. This group could claim significant progress in systems theory and in laying out the fundamentals for the then-upcoming cognitive science.

The interdisciplinary character of the Macy conferences on cybernetics becomes obvious when looking at a few specific topics of discussion. During the first conference self-regulation, neural networks and feedback mechanisms were presented, but also self-learning principles for computers and how to derive ethics from science. In the following year one topic was about communications among ant soldiers, while the third conference treated the topic of child psychology [34]. This widespread range of topics continued for the following conferences. A full list of conference topics can be seen at the website of the American Society for Cybernetics [34]. The scientific discussions formed a new worldview and the new cybernetics approach with a central element of communication and control mechanisms in complex systems. Already 1948, Wiener subtitled his book on cybernetics as Communication and Control in the Animal and the Machine [2] and hereby expressed the common core of this new approach [31].

4.3.2.4 Decoding the Basics of Self-organizing Systems

The contribution of Heinz von Foerster, an Austrian biophysicist, to the development of cybernetics is of major importance. He acted as co-organizer for the cybernetics conferences, was in charge of some of the proceedings and proposed naming the conference "cybernetics". During his entire career he was researching principles of interactions between humans, between humans and machines and between machines. This research was inspired by patterns of nature [31, 35, 41]. Foerster picked up many aspects of the cybernetics approach and developed them further. Since 1958 he held a research position at the University of Illinois and founded there the "Biological Computer Laboratory". This laboratory became a place of research for many scientists in the field of cybernetics, e.g. Ross Ashby [31, 42, 43].

The research focus of the Biological Computer Laboratory was on common structures of circular processes and their organization, for example cognition processes. Here, self-organization was of main interest in the first years. Heinz von Foerster discovered that order can arise not only from order but also from disorder. Among others, he described the example from nature, where single elements can be in a disordered state, and shaking results in an ordered arrangement—e.g. a crystalline structure [31, 43]. Heinz von Foerster transferred these observations on ordered states to several areas of life. He considered the phenomenon of self-organization in nature, technology and society to be of highest importance in science. Consequently, in 1961 he successfully held a conference on self-organization titled "Principles of Self-Organization", which generated results of major importance for this scientific field. Stafford Beer, a management cyberneticist called this conference the most important event of his life. And this event with its groundbreaking findings was the reason for the quickly increasing interest in self-organization as a field of research. Heinz von Foerster's publications were translated into many languages and distributed worldwide [31, 41].

Von Foerster decoded the abilities of self-organizing systems. These findings support self-preservation and therefore are of great use in many fields, e.g. as a basis for successfully managing complex systems [44]. This permitted theoretical approaches towards the artificial creation of self-organizing systems. Self-organization is of great significance for many areas, one being management. One major question is whether self-organization can be initiated and when a system starts to organize itself. Specifically in the management domain, it is important to know if big companies or other social constructs (e.g. national states) organize themselves (in a directed manner) or if such organization happens unplanned and arbitrarily. Assuming that the intended self-organization happens, a subsequent question is if this organization is meaningful or not. In management science, major researchers like Stafford Beer grounded large parts of their management doctrines on the preliminary work of Heinz von Foerster [45].

Today, von Foerster's findings get widely applied in the field of economics and management of social systems. Researchers like Stafford Beer and Frederic Vester picked up von Foerster's work and made steps towards practical applications of cybernetics [31, 46].

4.3.2.5 The Bio-Cybernetics of Frederic Vester

Frederic Vester was born in 1925 and became a biochemist and an expert for ecology. He advocated for replacing the ordinary, linear thinking by systemic thinking in order to face complexity issues. Vester is known as the founder of biocybernetics and created the term networked thinking (German: Vernetztes Denken). One of his most important works was the development of the eight basic rules of biocybernetics, which—according to Vester—are the precondition for a system the be able to live. As well, Vester declared that most common mistakes that occur during system planning can be avoided if these rules are observed and networked thinking gets applied as the fundamental principle [47]. The rules are formulated on an abstract system level and therefore are applicable to a large variety of human and ecological systems:

- Rule 1: Negative feedback cycles must dominate over positive feedback.
- Rule 2: The functioning of the system must be independent of quantitative growth.
- Rule 3: The system must operate in a function-oriented, not a product-oriented manner.
- Rule 4: Exploiting existing forces in accordance with the ju-jitsu principle rather than fighting against them with the boxing method.
- Rule 5: Multiple use of products, functions, and organizational structures.
- Rule 6: Recycling: Using circular processes (material and resources).
- Rule 7: Symbiosis: Reciprocal use of differences in kind through link-ups and exchange.
- Rule 8: Biological design of products, processes, and forms of organization through feedback planning.

According to Vester, positive feedback is important for making things run by self-enforcement. Rule 1 sees negative feedback or self-regulation by loops as even more important, as this is required for bringing a system in a steady balance.

The second rule states that a system that is primarily based on growth cannot reach a long-term balance and is likely to irreversibly exceed critical values. Such behavior and associated consequences were the result of the World3 model and the associated publication The Limits to Growth (see Sect. 3.4.3 for further details).

The third rule addresses the requirement of "flexibility and adjustment" of systems. Vester mentions that "Systems capable of surviving are geared to their function, not to their product" and that "products often change rapidly, whereas functions remain the same for a long time".

Rule 4 says that the own energy should only be used for steering and control, and that "using existing forces benefits from current situations and promotes self-regulation". Vester uses the example of a jujutsu fighter (in contrast to a boxer), who ideally used his power only for turning the power of an opponent towards himself. Like the jujutsu fighter, successful systems should absorb external impact and apply it for their intended purpose.

Advantages of multiple uses are covered by rule 5. Multiple use reduces flow capacity, energy, material and information efforts while increasing the degree of interconnectivity.

The principle of recycling, as formulated in rule 6, declares that systems (and societies) have to reach a state without waste. This could be reached when inputs and outputs of systems get completely interlinked—as it is the case in natural systems. This principle helps to reduce the risk of irreversible effects in the system.

Nature is also indicated as an example for rule 7. "Symbioses replace 'short-sighted exploitation' by 'stable cooperation'. The ecological and economic advantage of symbiosis is that, in leading to substantial savings of raw material, energy, and transport for all concerned, it takes pressure off the environment. But symbiosis calls for a certain smallness of scale and decentralized structures; it need a certain blending of functions [...] In other words, it calls for variety within a limited space."

Finally, rule 8 claims that products, functions and organizations "must conform to the structure of viable systems". Vester describes that "non-biological design ultimately fails to address the relevant demand and as such is produced without regard to the market. Yet countless planning disasters continue to result from decision-making processes that ignore this rule" [47].

The entire scientific work of Frederic Vester has been inspired by the idea of nature as a teacher. Vester thought that all answers to complex questions in developing technical appliances as well as when organizing human societies can be found by analogies to nature. Vester applied cybernetics as a method for systemic thinking and he developed a method-based software toolset for managing complex challenges [48]. And he brought his developments to many industry applications. As an author of popular science, Vester explained the enormous, mostly unused possibilities of biocybernetics as a future path to solutions to a broad audience. And he uncovered the risks of a lack of understanding complex systems and networked thinking.

4.3.2.6 Management-Cybernetics by Stafford Beer

The findings of Heinz von Foerster permitted, for the first time, to organize real complex systems without tremendous reduction or simplification. Stafford Beer was the first one to use these new opportunities for applying cybernetics to the field of management [49]. The mathematician, psychologist and philosopher was born in 1929. His lifelong scientific topic was about the effective organization of complex systems. This challenge was what Beer saw as a key factor for mankind in the future. Beer developed models and methods for managing the complexity of life, and called this work management cybernetics.

Stafford Beer identified the need for innovative and creative solutions in the field of management science and organizational design for industry as well as government. When starting his work in management cybernetics, he had already acquired substantial management experience from practical work in industry and the military. Beer had founded and led the largest operations research group in the industry for UK's former leading steel company United Steel, a group comprising 70 specialists from different disciplines. He had worked on highly complex problems in industry and disposed an interdisciplinary scientific thinking [31, 50]. Because of his practical experience, Beer already identified in the early 1950s the potential of Norbert Wiener's cybernetics for solving complex problems in the field of management. Based on this awareness, he transferred cybernetics' findings into a new management approach. Beer presented the results of this work in his book Cybernetics and Management in 1959 [51]. In this publication he strikingly showed that the highly abstract findings in cybernetics are fundamentally necessary for the successful design of complex systems [31, 50].

Beer's developments were influenced not only by the basics of Norbert Wiener's cybernetics, but also the fundamentals of self-organizing systems by Heinz von Foerster represented highly important groundwork for his management cybernetics. In fact, Beer based a theory of modeling comprehensively complex systems on the principles of self-organization [31, 52].

Stafford Beer had the intention to make the laws of organization and management applicable and therefore designed the model called "Viable System Model" [45]. He explained this model and its application in detail in the two books Brain of the Firm and The Heart of Enterprise [53, 54]. The Viable System Model describes the elements, functions and interconnections he thought of as being initially required for the viability of systems. It can also be applied for structuring information and communication in a large variety of systems [55–57].

Late in his scientific career, Stafford Beer published the significant book Beyond Dispute, which described further findings of fundamental importance [58]. He presented an innovative solution approach for one of the most significant problems of organizations—utilization of knowledge distributed in the organization. The so-called "Team Syntegrity" method matches the speed and effectiveness of small teams with the power of integration inherent in large groups [58, 59].

Beer's distinguished contribution to complexity management is the application of cybernetics for solving complex functions of human coexisting and cohabiting in organized structures like enterprises, states or other social systems. Before him, cybernetics had been applied to technical challenges; Beer succeeded in transferring the systems thinking and methods, and with that significantly widening the scope for this scientific approach.

4.3.2.7 Breakthrough in the German-Speaking Area

Three decades after its introduction, cybernetics has been implemented into many scientific areas. Also, the cybernetics approach significantly supported and enabled technical inventions associated with complex systems. Cyberneticists of the second generation like Stafford Beer and Frederic Vester extended the fundamental approach to new disciplines like management cybernetics and biocybernetics. But still publicity was lacking, for example no teaching programs existed in universities. And so the proliferation of cybernetics by spreading knowledge to young academics and entrants did not happen.

Early (first generation) cyberneticists like Norbert Wiener, Warren McCulloch and Heinz von Foerster were highly skilled scientists. The same accounts for the subsequent (second) generation of cyberneticists like Stafford Beer and Frederic Vester, who adopted and expanded the initial approaches. These scientists primarily aimed at improving their own disciplines, e.g. mathematics, physics, biology or psychology. And when they reached their discipline's limits they started to cross borders—and in doing so became system scientists. For the discipline of cybernetics the next step had to be a broader adoption of this science. Therefore the fundamental knowledge and its application needed to be significantly simplified and made applicable. Thus, the new challenge was to overcome the constriction of cybernetics to a small group of experts only [31].

Prof. Hans Ulrich taught at the University of St. Gallen, Switzerland and was among the first ones who focused on bringing cybernetics into broad scientific application. Ulrich recognized the significance of providing cybernetics' findings in an applicable form to a large audience. In 1983, Ulrich initiated the "St. Gallener Forschungsgespräche" (scientific discussions) with the topic "Self-organization and Management of Social Systems". The objective of this conference was to bring a large number of European scientists in touch with the fundamentals of cybernetics [60]. Heinz von Foerster, one of the early and famous cyberneticists (who disposed impressive rhetoric skills), was persuaded by Ulrich to participate in the conference and thus contributed to its tremendous success. Many participants adopted cybernetics in their later work. Those scientists can be seen as the third generation of cyberneticists. This includes for example the Austrian economist Fredmund Malik, who became the director of the Management Zentrum St. Gallen in 1977, the Swiss economists Peter Gomez, Gilbert Probst and the German psychologist Dietrich Dörner. This generation of cyberneticists began formulating the principles and methods of cybernetics in simplified language that allowed for its application without the need for highly specific qualification [31, 61].

4.3.2.8 Status Quo and Outlook

Many great scientists drove the revolutionary approach of cybernetics from the possibilities of holistic system modeling and understanding to applications in fields like biology and management. Some of the most influential people in this development of cybernetics are Ludwig von Bertalanffy, who developed the General Systems Theory; Norbert Wiener, the creator of fundamental cybernetics; Heinz von Foerster, the mentor of Biocybernetics; and Stafford Beer, who conceived applications of cybernetics in the management field.

Complex systems and their control require specific rules and approaches—this can be the summarized findings of cybernetics in the last century. And those rules and approaches as formulated by Norbert Wiener apply in the same way for systems of inanimate and organic nature. None of the single scientific disciplines (mathematics, physics, biology etc.) by itself is able to model and explain the phenomenon of complexity. For this reason, cybernetics had to be designed as an interdisciplinary approach from the early beginning on. If a phenomenon gets observed in one specific discipline, cybernetics can bring it to a higher level of abstraction. This way is also becomes manageable outside of the originating discipline and represents a new object of research [2, 31].

Today, cybernetics is widely applied in many different fields. The development of the computer, business management, education science, automation and psychotherapy are only some examples, whose growth are hardly imaginable without the influence of cybernetics. Or at least their development would had been tremendously different from what we know today.

In this context it is an interesting observation that most people are not familiar with the approach given by the Control and Communication in the Animal and the Machine. In fact, not even the term cybernetics can be denominated as general knowledge. If people have some association with the word, then it is most likely with one of the derived terms like cyborg, cyberspace—and yes, also cybersex. While the implications of cybernetics in these terms is obvious and meaningful, it can only be speculated that there is no broader knowledge about it. One possible explication is that cybernetics is so heavily interwoven into the daily life that it is not specially noted and seen as a matter of course.

While cybernetics has been and still is applied in many technical and natural scientific applications, there is still much unused potential for organizational and societal design and improvement. One striking example could be the new appearance of wearable computer devices, specifically the augmented reality device Google Glass. There are still many technical challenges for this man-machine-interaction device that have to be solved, e.g. adequate energy supply, possibilities of intuitive user input and context-based user support. But the even larger question is how millions of those devices in daily application will impact society. Challenges to be solved are for example, the impact and limitations of data tracking, personal privacy and information-sharing behavior. Organizational design and new management approaches for the digital age are challenges on the doorstep—and for those challenges cybernetics provides powerful fundamentals.

4.3.3 Operations Research

The approach of operations research came up around 1940 from a military context. It arose from a group of British scientists (later joined by American scientist), who developed a scientific approach for investigating military operations during Second World War [62].

One central idea when developing operations research was to consider the entire system when solving a problem (which appears within this system) [63]. Therefore, the approach included intensive interdisciplinary collaboration of military and scientific experts. Operations research did not aim at gathering new scientific findings, but solving specific challenges, which could also be characterized by significant dynamics. As well, problem solutions included the readiness for applicability, thus solutions were not only given on a theoretic level, but as an instruction for implementation. Up to the present day, this understanding of operations research is still one of its main characteristics.

Operations research arose during the Second World War. Other than cybernetics, operations research cannot be traced back to a single founder or name patron. The breeding grounds for this new approach were in wartime Britain shortly after the start of the Second World War. Great Britain suffered severe defeats in the first year of the Second World War. The beginning of German air strikes in 1940 aggravated the problematic situation. Soon Great Britain was largely depending on supply ships from the United States. But even those ships were soon under massive attack from the German submarines and suffered significant losses. Consequently, one of the most urgent challenges for the Allied Forces was to detect and destroy German submarines [28].

Because of this state of emergency, the English command was forced into an extensive search for solutions to this challenge. They encountered an interdisciplinary group consisting of scientific and military employees, which had already been founded in 1937. The group's mission was to investigate the optimal layout of a radar control system for the British military [64, 65]. Three years later the group could already claim some success in early radar detection of hostile airplanes in the Battle of Britain [66]. In 1940, this success resulted in building up similar groups for the British air force, army and navy [65]. One exemplary follow-up project was the analysis of air force attacks against submarines, which was executed by the Navy Operations Research team. A statistical analysis of previous strikes against submarines led to the suggestion of changing the depth adjustment of depth charges. This action was reflected in a significant increase of hostile submarines sink rates [64].

The physicist Patrick Maynard Stuart Blackett was the head of a famous interdisciplinary team and convinced the British leadership of the necessity of a scientific approach for complex operations. For this reason, he is often mentioned as the founding analyst of operations research [67]. In 1942, similar groups were started in the United States, which also generated a new research field of mathematical strategies for the planning and optimization of military operations.

In the following years, operations research was increasingly applied to many problems of strategic relevance that the Allied Forces were confronted with. One famous example of applications became the protection of US convoys against German submarine attacks. Those convoys were shipping supplies over the Atlantic Ocean under the constraint of extremely limited resources for each military operation. The challenge was about optimizing the chances of success for the convoys (meaning to minimize losses of ships by submarine attacks) by disposing the limited defense resources.

The successful, efficient solution for the configuration of ship convoys was of significant importance for the entire course of the Second World War [64, 65]. And the formal collaboration between science and the military made a substantial contribution to the victory of the Allied Forces in the naval war against the German navy [68].

This impressive success of operations research applications during the Second World War led to intensive subsequent use and further development of this approach in the US and Great Britain. After the war, experts left their military field of work in order to migrate into economics and proceed and extend their applications of operations research. The application of electronic calculators for solving scientific and administrative challenges was a pioneering work in the field of operations research, and this work was powered by the findings gained during the Second World War [68].

At least since the invention of the simplex algorithm for solving linear programming in 1947, operations research became noted as a scientific approach [66, 69]. In close connection with the development of computers in the 1950s it finally became accepted as an independent scientific discipline. One of the drivers of this movement was Saul Gass, working at the University of Maryland [70].

Especially in the United States the successful collaboration between scientists and military personnel has been cultivated ever since then. Until today, so-called "think tanks" serve as advisors for the political government [68]. One important institution is the RAND Corporation, founded at the end of the Second World War in the environment of the US Air Force [71]. During the Cold War, the main objective of RAND was to generate strategic military scenarios for the US government. But also socio-scientific research has been conducted by this company, which had famous employees like John von Neumann and Donald Rumsfeld. Besides solutions to security and political questions, also many other important questions were on the agenda of RAND. One exemplary project from recent times is the increasing obesity of the US population [65, 68]. Many companies made successfully use of operations research for solving complex challenges. And operations research became established in several scientific fields, e.g. engineering and economics.

Since the 1960s, operations research also became mentioned in technical literature in Germany, beginning with translations of publications by Curchman, Vazsonyi and Sasieni. In the beginning, distinguished German expressions were created, e.g. Unternehmensforschung, Optimalplanung, Planungsforschung, Planungsrechnung or mathematische Entscheidungsfindung. However, none of them really gained traction and operations research became the common term in use [70]. Operations research became introduced to universities and lectures as an interdisciplinary approach of applied science—and finally as a research field of its own. For example, the Rheinisch-Westfälische Technische Hochschule in Aachen (RWTH Aachen) possesses a chair for operations research. Today, operations research is widely accepted in industry as a method for planning and decision making.

The development and application of operations research has been influenced significantly by the advent of electronic data processing systems. Today, it is possible to tackle complex problems with software programs, assuming that one is capable of the operations research basics. Especially in the field of simulation, electronic data processing provides tremendous possibilities. Operations research gained the widest penetration in the field of economics, where the term management science was established for this kind of approach [72].

Organizations and conferences carry the knowledge and carry out new development in the field of operations research. On the international level IFORS (International Federation of Operations Research Societies), INFORMS (Institute for Operations Research and the Management Sciences) and SIAM (Society of Industrial and Applied Mathematics) should be mentioned. IFORS further comprises continental and country-based sub-organizations; for Germany this is the GOR (Gesellschaft für Operations Research e.V.), a lively community that organizes many conferences worldwide.

4.3.4   Systems Engineering

It is rather difficult to identify the origin of systems engineering, as the basic concept is so universally valid. Marvel mentions that one could see the earliest systems engineering approach in the first house construction that required different specialists for executing the construction [73]. Consequently, collaboration between those specialists was useful and required planning and communication, which is challenging, because of differences in terminology in different disciplines. A central communicator, planner and organizer became necessary to manage the project. Brill states that "Based on the impressive civil engineering and other projects of ancient cultures, it is reasonable to assume that in today's terms these ancient engineers would be regarded as systems engineers. Also, few would argue that ancient philosophers such as Aristotle possessed the attributes often ascribed to individuals that take a systems or wholistic approach to problem solving" [74].

The continuous technological advance was boosted in the age of industrialization, which brought innovations like the steam engine by James Watt. In accordance with this advance, also expert knowledge and technical terminology in each sub-discipline increased dramatically. This led to linguistic differences which impeded communication across disciplines. And finally, this fostered the barriers between disciplines and impeded collaboration of specialists from different fields. An improvement to this situation was to install a coordinator for managing the collaboration between the specialists of different fields [73].

During and shortly after the Second World War significant research was done in the field of systems thinking, resulting in new approaches and methods for managing complexity. While this research was targeted at specific applications, e.g. anti-aircraft cannons, the scientists involved were mostly theorists with little relation to complexity issues as they occur in evolving enterprises. But such enterprises were lacking management approaches for the development and management of complex technical systems.

At the end of the 1940s, US-containment politics brought up war scenarios between the two new superpowers of the USA and the Soviet Union. The US government felt responsible for containing the spread of communism in the world, as it was enforced by the Soviet Union. In the shadow of the mobilization for the Korean War in 1950, the demand for application-oriented research became more explicit. The new systems engineering approach was meant to fulfill these demands. Engineers with a holistic thinking in the sense of cybernetics linked established methods from communications engineering with cybernetics. This formed the methodical framework for dealing with engineering disciplines, as the manifold aspects of cybernetics became integrated into engineering—which meant a contrast to the trend of differentiation of technical disciplines [30].

Systems engineers thought of themselves as pioneers who would change engineering sciences with interdisciplinary approaches. An engineer should become a technical generalist who can overcome the frontiers between disciplines. This movement asked for a united engineering science, according to the large-scale military projects with interdisciplinary teams executed during the Second World War [30].

As already mentioned, it is not possible to identify the exact time of systems engineering's origin. At least the term arose from work done at Bell Labs in the US (see Sect. 3.4.2). In 1940 the term was used first in the context of new weapons development. The fundamental approach, however, was already in use at Bell Labs since approximately 1900. The US Department of Defense used systems engineering in 1940 for developing missiles and the defenses against them. Then in 1946 the RAND Corporation applied system analysis as a part of systems engineering.

Until the 1960s, most relevant work in systems engineering was motivated by the demands of the Cold War, e.g. network analysis in communication and transport, design of electronic system components, data processing, industrial automation, process control and development of weapon systems [30].

Early systems engineering could apply modern procedures and methods of probability theory, statistics, game theory, linear programming, information theory, cybernetics et cetera. In general, all these areas were highly theoretical and comprised significant amounts of mathematics. This changed the ideal perception of an engineer from being occupied with experiments, technical drawings or practical assembly work to instead work on the mathematical modeling of components, system characteristics and abstract computations. Thus, early systems engineering implied the mathematization of engineering. However, the initial positive acceptance of this development was later seen as a fundamental problem in engineering science [30].

Brill published a timeline indicating the development of systems engineering from 1950 until 1995 based on the work published by some of the key contributors. An adapted version of this timeline is depicted in Fig. 4.6. Statements cited by Brill from these key contributors are aggregated in Table 4.1.

According to Hall, "probably the first formal attempt to teach systems engineering was made in 1950 at the Massachusetts Institute of Technology by Mr. G. W. Gilman, then Director of Systems Engineering at Bell Laboratories, Inc." [75] (according to [74]). Then, "Goode and Machol of the University of Michigan published Systems Engineering in 1957 in which they observed a phenomena of systems thinking and approaches to designing equipment" [76, 87].

Hall published the book "A methodology for Systems Engineering" in 1962, which described a concept for Systems Engineering. In 1989, Hall published the book Metasystems Methodology, in where he laid out details about the approach he called system methodology. Hall highlighted the importance of these new methodologies for dealing with complex problems.

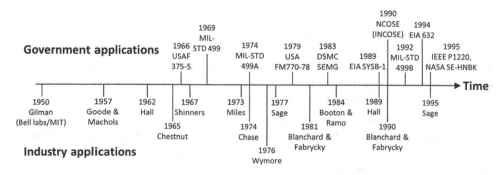

Fig. 4.6 Timeline of Systems Engineering Contributors (adapted from [74])

Table 4.1 Key publications in Systems Engineering from 1950 to 1995 (according to [74])

Author/Institution	Title/Description
Gilman	First teaching of Systems Engineering (according to [75])
Goode and Machol [76]	"for more than a decade, engineers and administrators have witnessed the emergence of a broadening approach to the problem of designing equipment. This phenomenon has been poorly understood and loosely described. It has been called systems design, systems analysis, and often the systems approach." "systems design entails many things: a new set of tools, a new classification of parts, an organized approach albeit seemingly chaotic, and a team of workers. The time is ripe to weld these many things together"
Hall [75, 77]	Concept of systems engineering consisting of three elements: systems engineering and its multifaceted definition, three divisions of the environment (physical/technical, business/economic, social), considering the needs of customers and how to fulfill these needs. Insights into systems methodology (SM), defined as: "A body of thought, theory, procedures, and specific methods applicable [...] to most if not all, 'complex problems'. The subject is enormously important, because it integrates all of the ways that man has to improve his world, and it adds a few to fill some voids, as if to glue together the parts into a new unified synthesis, with the power to cope with increasingly complex problems."
Chestnut [78]	Guidance on how to formulate the problem and acquire requirements
Shinners [79]	"seven general procedures are involved in engineering an overall large complex system"; "this logical unified systems engineering procedure is in reality a feedback process". "The best advice for solving a system-oriented problem is first to understand the problem. The systems engineer must fully determine the overall system requirements and objectives, and at the same time, fully understand the constraints imposed."
Miles [80]	A lecture series by experts on "Systems Concepts for the Private and Public Sectors", held at the California Institute of Technology in 1971, edited by Miles. Description of a six-step approach towards system management.
Chase [81]	General description of Systems Engineering. "there are tremendous language difficulties to be overcome in effectively communicating systems concepts and in describing the systems approach".
Wymore [82]	"an interdisciplinary team must be the nucleus of system design". "system analysis is driven by three imperatives: modeling human behavior, dealing with complexity and largeness-of-scale, and dealing with a dynamic technology".
Sage [83, 84]	Comprehensive overview of engineering of large-scale systems; explications on the topics of system methodology, design and management, system quality assurance, configuration management, audits, reviews, standards, system integration. "systems engineering is the management technology that controls a total

(continued)

Table 4.1 (continued)

Author/Institution	Title/Description
	lifecycle process, which involves and which results in the definition, development, and deployment of a system that is of high quality, trustworthy, and cost effective in meeting user needs". Introduction of a three-level systems engineering approach (structure, function, purpose).
Blanchard and Fabrycky [85]	"System-life-cycle-engineering" defined as "starting with the initial identification of a need and encompassing the phases (or functions) of: planning; research; design; production or construction; evaluation; consumer use; field support; and ultimate product phaseout". "systems engineers must discipline themselves to think in terms of the system-life-cycle to ensure that all aspects of the system are considered".
Booton and Ramo 1984 [86]	"large scale attention to modern systems engineering occurred in the post-war (WW II) developments of ground-to-ground, ground-to-air, and air-to-air missile systems, where technologies involved included communications, radar, controls, aerodynamics, structures, and propulsion".

Chestnut as well as Shinners pointed at the importance of early stages in the systems engineering process, giving guidance on formulating the problem, acquiring the system requirements and understanding the general problem that has to be tackled [78, 79].

In 1973, Miles published his edited version of a lecture series on system concepts, which was held by a group of experts at the California Institute of Technology. In this publication, Miles also presented a six-step approach for system management [80]. One year later, Chase provided a general description of systems engineering, which he linked to communication systems concepts [81]. Sage also contributed a comprehensive overview of methodical engineering of large-scale systems [84]. And Wymore strengthened the importance of an interdisciplinary team as the "nucleus of system design" [82].

Sage as well as Blanchard and Fabrycky mention the life cycle perspective of systems engineering. Sage mentions that "systems engineering is the management technology that controls a total lifecycle process", and Blanchard and Fabrycky define "system-life-cycle-engineering" by the process steps from identifying the need until the "ultimate product phaseout" [83, 85] (according to [74]). And in 1984, Booton and Ramo describe in their retrospective publication "The Development of Systems Engineering" that (mainly military) system demands drove the need for modern systems engineering.

Because of systems engineering's relevance, industry and governmental organizations developed their own standards and handbooks for this approach on solving complex challenges. The earliest of such documents was written by the United States Air Force (USAF) in 1966, when they published their Handbook 375-5, which contains exact descriptions of the systems engineering process [88]. Several other documents followed with the intention to improve the standardized guidance on systems engineering [89–94].

Professional associations like the Electronic Industries Association (EIA), IEEE and NCOSE/INCOSE started publishing their own standards in the 1990s [95–97]. Currently, INCOSE published the fourth version of their Systems Engineering Handbook [98], which has become a quasi-standard in many parts of industry.

According to Gorod et al., Keating et al. mention "shortcomings in the ability to deal with difficulties generated by increasingly complex and interrelated system of systems" represented the next challenge in the field of systems engineering. "There was a need for a discipline that focused on the engineering of multiple integrated complex systems [99] (according to [100]). "Today, we refer to this as SoSE [99]. So, system-of-systems engineering is a further development of systems engineering in recent times. A historical development of this approach is presented by Gorod et al., who also created a timeline of selected contributors similar to the timeline of systems engineering history published by Brill for the timeframe of 1950–1995 [74, 99]. This timeline covering the years 1990–2008 is displayed in Fig. 4.7. Selected statements of the authors appearing in this timeline are aggregated in Table 4.2.

In the early 1990s several publications introduced the term system-of-systems and provided initial definitions [101–104]. Owens as well as Manthorpe highlight the importance of this approach for military applications [105, 106]. Maier points to the fact that a SoS is an assembly of systems with operational and managerial independence of its components [107, 108]. Kotov then "was one of the first scientists to attempt to model and synthesize SoS", Luskasik applied SoS to the educational context and Pei contributed the concept of system-of-systems integration in a military context [109–111]. Next, Carlock and Fenton enhanced SoS to "enterprise systems of systems engineering" by integration of enterprise activities like strategic planning and investment analysis [112]. Several authors focus on a framework for SoSE, and Keating et al. published guidelines for SoSE phases based on a comparative study of systems engineering and system of systems engineering [100, 114, 115]. Other authors aggregated definitions for a system of systems based on various sources [117, 119]. And in 2008 two books were

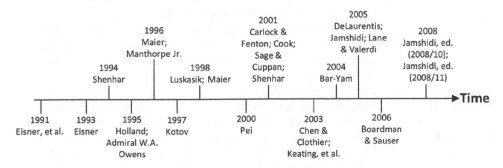

Fig. 4.7 Modern history of system of systems (adapted from [99])

Table 4.2 Key publications in Systems-of-Systems Engineering 1990–2008 [99]

Author/Institution	Title/Description
Eisner, et al. [101]	Definition of a system-of-systems as "A set of several independently acquired systems, each under a nominal systems engineering process; these systems are interdependent and form in their combined operation a multifunctional solution to an overall coherent mission. The optimization of each system does not guarantee the optimization of the overall system of systems"
Eisner [102]	Introduction of the modern term SoS
Shenhar [103]	Definition of a system-of-systems as "A large widespread collection or network of systems functioning together to achieve a common purpose". "Shenhar was one of the first to describe SoS as a network of systems functioning together to achieve a common purpose."
Holland [104]	"Holland proposed to study SoS as an artificial complex adaptive system that persistently changes through self-organization with the assistance of local governing rules to adapt to increasing complexities."
Owens [105]	Owens "introduce[d] the concept of SoS and highlight the importance of its development in the military".
Manthorpe [106]	"In relation to joint warfighting, system of systems is concerned with interoperability and synergism of Command, Control, Computers, Communications, and Information (C4I) and Intelligence, Surveillance, and Reconnaissance (ISR) Systems."
Maier [107, 108]	Maier "proposed for the first time to use the characterization approach to distinguish 'monolithic' systems from SoS. These characteristics include 'operational independence of the elements, managerial independence of the elements, evolutionary development, emergent behavior, and geographical distribution'". "A system-of-systems is an assemblage of components which individually may be regarded as systems, and which possesses two additional properties: Operational Independence of the Components [and] Managerial Independence of the Components [. . .]."
Kotov [109]	"Systems of systems are large scale concurrent and distributed systems that are comprised of complex systems". Kotov "was one of the first scientists to attempt to model and synthesize SoS".
Luskasik [110]	"SoSE involves the integration of systems of systems that ultimately contribute to evolution of the social infrastructure." "Luskasik attempted to apply SoS approach in the educational context".
Pei [111]	Introduction of "a new concept of 'system-of-systems integration' (SOSI) which gave the ability 'to pursue development, integration, interoperability, and optimization of systems' to reach better results in 'future battlefield scenarios'".
Carlock and Fenton [112]	Suggestion on joining "traditional systems engineering activities with enterprise activities of strategic planning and investment analysis"; introduced the term "enterprise systems of systems engineering"
Cook [113]	"Cook [. . .] described a distinction between 'monolithic' systems and SoS based on 'system attributes and acquisition approaches'. Constituent systems of SoS are acquired through separate processes."

(continued)

Table 4.2 (continued)

Author/Institution	Title/Description
Sage and Cuppan [114]	Proposition of "principles of 'new federalism' to provide a framework for the SoSE"
Keating et al. [100]	Comparative study of SE and SoSE; Provision of guidelines for several key phases such as "design, deployment, operation, and transformation of SoS"
Chen and Clothier [115]	Indicating the "need for a SoSE framework"; suggestion for "advancing SE practices beyond traditional project level to focus on "organizational context"
Bar-Yam et al. [116]	Suggestion to add "characteristics as opposed to definitions to provide a more comprehensive view of SoS"
Jamshidi [117]	"Definitional approach to SoS by collecting different definitions from various fields"
Lane and Valerdi [118]	Identification of "other universally known network-centric systems as examples of collaborative SoS (i.e., the internet, global communication networks, etc.)". analysis of "SoS definitions and concepts in the 'cost models' context"
Boardman and Sauser [119]	"outlined the characterization approach to SoS"; Identification of "patterns and differences in over 40 SoS definitions"; "comprehensive overview of five distinguishing characteristics of SoS", the characteristics are. autonomy, belonging, connectivity, diversity, and emergence.
Jamshidi [120, 121]	"first two books dedicated to SoS", covering "a wide variety of SoS topics"

published with a special focus on SoS, covering many aspects of this enhanced systems engineering approach [120, 121].

Probably the most famous example of systems engineering application is the NASA Apollo Program in the 1960s [122]. In the context of the space race with the Soviet Union, President John F. Kennedy held his historic "Moon Shot Speech" in Congress on October 25 1961, stating: "First, I believe that this nation should commit itself to achieving the goal, before this decade is out, of landing a man on the moon and returning him safely to the Earth." This extremely complex task was divided into manageable partial tasks and worked on by hundreds of agencies, authorities and enterprises involved in the Apollo Program (Fig. 4.8). And all partial tasks finally contributed to an integrated, holistic solution.

From its beginnings on, the application of systems engineering was closely linked with the aeronautics and astronautics industry and software for technical appliances. Besides the aforementioned Apollo Program, systems engineering became also applied and methodically improved in the development of the US Space Shuttle and in the European aeronautics program, e.g. for the development of the Ariane rockets. Nowadays, many universities offer systems engineering as a field of study, often linked to aeronautics or astronautics departments.

In 1990, the International Council of Systems Engineering (INCOSE) was founded as an association specializing in the teaching, application and research in systems engineering. Since then, national chapters have been founded all over the world with great success. Besides INCOSE, several professional institutions, e.g. the Institute of Electrical and

Fig. 4.8 President John
F. Kennedy in his historic speech
to the Congress, May 25, 1961,
NASA (Great Images in NASA
Description), via Wikimedia
Commons

Electronics Engineers (IEEE) and the American Institute of Aeronautics and Astronautics
(AIAA) run their own departments with a focus on systems engineering [123, 124]. And
enterprises like Boeing and EADS operate systems engineering programs with own
conferences and advanced training.

System Engineering vs. Engineering Design
When comparing approaches and methods of systems engineering with those taught in
engineering design, large overlaps can be found. While it is difficult to draw a clear line
between the disciplines, differences can be found in their historical origin. As introduced in
this chapter, systems engineering was based on the holistic thinking of cybernetics and
represented a technical application of the thus far theoretical methods. Statements like
Kennedy's famous "Moon Shot Speech" represented top-down challenges, starting with a
complex holistic target, which had to be broken down into manageable tasks.

Engineering design has its background in a systematic approach toward the engineering
of products. Beginning with purely mechanical devices, step by step more disciplines
became integrated for realizing more complex product functionalities. Electrical
components became integrated, then electronics and software, which led to the definition
of mechatronic products. Each discipline brought more complexity to the system, communication, interface and integration challenges. This development of engineering design can
be seen as a bottom-up process, integrating more and more disciplines and bringing more
and more complexity into the system. Finally, mechatronic systems like an automobile
require a systems engineering approach.

Current Importance of Systems Engineering
Systems engineering concepts quickly received attention after successful application in the
Apollo Program in the United States and additional aeronautics applications in Europe.
Nowadays, its integration into large-scale technical processes is mandatory and regulated
by several norms. Application of systems engineering results in faster problem solving and
higher product quality, due to the fact that the amount of possible failures of an entity is
generally higher than the possibilities to fail for its single parts [125].

Systems engineering became a synonym for a systematic approach towards the entire product design process of complex hardware and software systems. Also successful application to socio-economic systems have been documented [126]. Nowadays, systems engineering gets not only applied to extremely large projects, but significant effort is also put into transferring systems engineering methods into medium-sized businesses with their smaller scope of applications.

Enterprise Architecture and Architecture Frameworks
The design and application of architecture frameworks has become one of the major approaches toward managing the complexity of systems. The origin of this approach can be traced back to the works of P. Duane Walker in the late 1960s. As the director of architecture at IBM, Walker "had established an enterprise analysis-oriented planning tool called Business Systems Planning (BSP)" [127]. Beginning in the early 1980s, Walker's student John Zachman carried the BSP approach forward and in 1987 published his findings in the article titled "A framework for information systems architecture". Zachman stated that "with increasing size and complexity of the implementations of information systems, it is necessary to use some logical construct (or architecture) for defining of the components of the system" [128]. Based on analogies to other disciplines like (building) architecture and airplane design he elaborated a framework for information architecture, which has been the initial point for many frameworks to follow.

Zachman's conclusions from his research were that "There is not an information systems architecture, but a set of them!" and that "we are having difficulties communicating with one another about information systems architecture, because a set of architectural representations exists, instead of a single architecture. One is not right and another wrong. The architectures are different. They are additive and complementary" [128]. Zachman's information systems framework classifies types of architecture description by user perspectives (users are e.g. the owner, designer and builder) and by purposes (description of material, function and location). Each combination of perspectives and purposes (typically displayed in a grid) results in a different architecture representation (view).

While Zachman developed his framework specifically for application to information systems, he also provided model associations on a generic level. Material, function and location descriptions represent what the system uses to operate, and how and where the system operates. In later works, Zachman added the perspectives of people, time and motivation, which describe who, when and why the system operates [127]. Today, the education and consulting firm Zachman International states that "there is substantial evidence to establish that the Zachman Framework™ is the fundamental structure for Enterprise Architecture and thereby yields the total set of descriptive representations relevant for describing an Enterprise" [127].

Beginning in the late 1980s, several other architecture frameworks were developed, which were significantly based on Zachman's prior work. In 1986, the United States Department of Defense started working on their enterprise architecture reference model

Technical Architecture Framework for Information Management (TAFIM), which served as the basis for The Open Group Architecture Framework (TOGAF) published in 1995 [129]. The Federal Enterprise Architecture Framework (FEAF) resulted from the Clinger-Cohen Act in 1996, a US Congress initiative that aimed at improving the way of managing and investing in IT resources in federal agencies. The perhaps most popular architecture framework is the Department of Defense Architecture Framework (DoDAF), which was published as version 1.0 in August 2003 (current version is 2.02), but has its roots and precursor frameworks in the 1990s. "The Department of Defense Architecture Framework (DoDAF), Version 2.0 is the overarching, comprehensive framework and conceptual model enabling the development of architectures to facilitate the ability of Department of Defense (DoD) managers at all levels to make key decisions more effectively through organized information sharing across the Department, Joint Capability Areas (JCAs), Mission, Component, and Program boundaries" [130].

Within a couple of years, architecture frameworks became extremely popular and distinctive frameworks were created for many different use cases and organizations, e.g. AUTOSAR as a standard for the automotive E/E architecture [131]. Already in 2003, Jaap Schekkerman published his book "How to Survive in the Jungle of Enterprise Architecture Frameworks" [132]. A comprehensive and current survey of architecture frameworks can be found at the web site pertaining to the standards document "Systems and software engineering—Architecture description ISO/IEC/IEEE 42010" [133].

4.3.5   System Dynamics

Jay Wright Forrester was born in 1918 on a farm in Nebraska. In one of his later publications he mentioned that the practical and technical work at the farm during his youth did form his later thinking [134]. Forrester got a degree in electrical engineering, became a management expert, and researched and taught at MIT until he retired. He is known as one of the intellectual fathers of the computer and a pioneer in system dynamics. In the beginning called industrial dynamics, system dynamics represents a method for the holistic analysis and modeling of complex, dynamic systems [135].

After the Second World War, Forrester worked at "Whirlwind", a flight simulator project for the US military forces. The objectives of the project were the development of a digital computer, creation of virtual reality and the investigation of the interaction between this virtual reality and humans. According to the cybernetics' claim for a universal science, Forrester also aimed for a broad spectrum of application for these computers.

The computer, which resulted from Project Whirlwind was applied to the continental defense system of the United States and coordinated incoming data from radar stations, airplanes and other military objects. This application confirmed Forrester in his opinion that large projects require powerful computers, which work in real-time mode [136]. Forrester pursued this objective for the next decades.

One of the outcomes from Project Whirlwind was a new type of memory chip, developed by Forrester. This technological advance had significant impact on the further development of computers in general. Another side note worth mentioning is that the first animation in the history of computer graphics can be traced back to Jay W. Forrester [68, 135].

In 1956, Forrester moved to the MIT Sloan School of Management. His main objective in this position was to integrate economic science with engineering science in research and teaching. He planned to work in the field of operations research too, but was concerned about the missing relevance for the field of management; for example the success and failure of companies were not considered by operations research applications. As the applicability of research findings was of major importance for Forrester, he did not further pursue this approach [135, 137]. In 1957, Forrester founded the System Dynamics Group and laid the cornerstone for the system dynamics approach [134].

In Forrester's thinking, enterprises represent dynamic systems, wherein production parameters like staff, capital, commodities and machines are not static values, but parameters that are continuously developing. Consequently, this requires dynamic and real-time focused management [68]. As Forrester initially aimed with his method towards industrial applications, he called his method industrial dynamics.

Industrial dynamics still represents the standard work about system dynamics, published by Forrester in 1958 [138]. This work was preceded by a General Electric assignment, which asked for the origins of problems in capacity use in one factory. Forrester modeled the problem situation and simulated the temporal development by the use of a computer. His findings showed an oscillating structure of an unstable system. The consequence of this oscillation was that despite an unchanged order situation, unstable employment resulted because of existing policies. Once the system structure was understood, it became much easier to interpret similar system structures. While in the beginning only technical systems were modeled, now also social system structures moved into the focus of Forrester's approach. Consequentially, the term industrial dynamics was changed to system dynamics [137].

In 1971, Forrester published the book World Dynamics, which describes in detail his "world model" for simulations [139]. This model became adopted for the project "Limits to Growth", which was led by Forrester and was noticed by a broad audience. Forrester's "world model" became a topic of intense discussion by scientists at the time. One controversial aspect was the application of mathematical, computer-supported simulation to societal challenges. Compared to empirical approaches, is became questioned if such simulations—without empirical evaluation—can be seen as scientific procedures [68].

Similar to many other system science developments of the same time, Forrester's scientific approaches emerged from the cooperation between military and academia during World War II, the Korean War and until the time of the Cold War. However, Forrester did not only apply his findings in the military context, but also brought system dynamics into many civil applications. Due to the computational demands of his simulation approaches,

Forrester was convinced that solving complex challenges requires mainframe computers [68].

Some system dynamics experts like Wolstenholme and Coyle aimed at spreading the method to a broader audience, especially to students in higher education. They thought that this can be realized best if methods and visualizations were kept simple, without the use of simulation techniques. Nevertheless, they propagated system dynamics as a method comprising two different modeling phases, the qualitative one and the quantitative one, and both can be applied for investigating problems [137, 140, 141].

J. Sterman thought that only simulation allows for understanding systems correctly and refused the sole application of qualitative system dynamics models [142]. J. Forrester did not change his methodical approach several years after his first publication, but enlarged the spectrum of application. In 1969, he published the book Urban Dynamics, investigating the interdependencies between governmental policies and poverty in cities [143].

Today's Significance of System Dynamics
In the early years after its development, system dynamics required large efforts for creating the initial models, which then build the basis for the simulation part. The model-building was executed by experts only. And computer-supported simulation required special software like DYNAMO [134, 139]. Therefore, system dynamics was applied on large-scale projects, where the efforts were affordable. This constraint was overcome when computer systems became more powerful, cheaper and consequentially more widely distributed. This development was accompanied by increasingly better software applications, which also became easier to apply. With modern hardware and software, the effort required for model creation and simulation is much smaller, hence making the application of system dynamics also profitable for small projects. Today, many simulation tasks can be run by a single user [144].

Nowadays, system dynamics gains worldwide interest, e.g. indicated by the increasing popularity of the "Systems Dynamics Society". Founded by academics in the 1980s, this association takes charge of an international exchange of opinions and findings in the field on yearly conferences and in interest groups. Since 1985 the Systems Dynamics Society publishes the journal System Dynamics Review [145].

In Germany the "Deutsche Gesellschaft für System Dynamics e.V. (DGSD)" acts as the local chapter of the System Dynamics Society with similar objectives. Prof. Gert von Kortzfleisch (1921–2007) was the pioneer of teaching system dynamics and its application in Germany at the University of Mannheim. Von Kortzfleisch worked together with Jay Forrester at MIT in 1968. Two of his co-workers, Peter Milling and Erich Zahn, worked in Dennis Meadows's group at MIT in the early 1970s and contributed to the model, which represented the basis for the famous publication The Limits to Growth [144, 146]. Over time, system dynamics became a commonly applied technique when analyzing complex challenges in the academic field. It is taught in many academic training programs and is a building block in many research projects. And an increasing number of enterprises make use of system dynamics for decision-making in case of strategic challenges [144].

According to Sterman, system dynamics gets widely applied in fields like society, politics, economics and science [142]. Examples that reached some popularity are recommendations for action for enterprises of the air traffic industry concerning the cyclical development of their industry and improvements of material flows in the enterprises' supply chain [144]. Rüffer describes the application of system dynamics for supplementing project management [147]. And Schwarz & Ewaldt describe the assessment of technological evolution by system dynamics modeling [148]. Besides problem-focused applications, system dynamics also gets applied for management training [149]. Based on a reference model, managers can analyze, validate and improve their own mental models [150, 151].

Information dynamics represents a specific development of system dynamics. It is based on the assumption that information is of major significance for system behavior, as this behavior results from interactions (information exchange) between system elements. Findings show that efficient system control can be reached with optimized information processing [152]. In this context, the "Beer Game" is a simulation game designed as a "flight simulator for management education" by Sterman, based on a preceding role game [153, 154]. "The game was developed by Sloan's System Dynamics Group in the early 1960s as part of Jay Forrester's research on industrial dynamics. It has been played all over the world by thousands of people ranging from high school students to chief executive officers and government officials" [153].

4.3.6 Game Theory

Game theory focuses on the search for optimal decisions in situations with multiple actors. These actors are interacting, acting rationally and are all aware of this fact. Rational thinking means that all actors aim at maximizing their own benefit [155]. The application of game theory analysis tools provides insights to popular situations, e.g. in politics or economics. One well-known example is the prisoner's dilemma, modeling the behavior of people in the context of self-interest and possible collaboration. Furthermore, game theory identifies structural interdependencies, which then can be verified in subsequent experiments. And it can provide acting guidelines for complex situations [155, 156].

Game theory has been applied to a wide variety of domains, with its first use cases in economics. Already in 1944, John von Neumann and Oskar Morgenstern applied their theoretic approach to economics in their groundbreaking book Theory of Games and Economic Behavior [157]. Besides economic applications like the consideration of market situations, game theory got also intensely applied to biological challenges and political, social and military problems [155]. Operations research applications also make use of game theory.

Originally, game theory served as the mathematical descriptions of decision behavior in parlor games like chess and checkers. Already in 1921, Émile Borel published an article with the title "La théorie du jeu et les équations intégrales à noyau symétrique" (English:

Game theory and integral equations with symmetric kernel) [158]. Game theory as a specific scientific approach was then introduced in 1944 with the book authored by von Neumann and Morgenstern [155]. During the Second World War, von Neumann worked as consultant for the US military, then joined the Manhattan Project and became a member of the RAND Corporation, where he applied game theory in strategic mental games [159].

After the Second World War, early applications of game theory were carried out in theoretical economics and strategic and tactical questions of warfare and military planning [156]. During the Cold War, game theory came to be applied on both political sides [159].

At the first level of its development, game theory could only be applied to zero-sum games, using the so-called minimax theorem. Those games are characterized by the fact that the total benefits and losses of all participating players equals zero (or a constant amount, which is known in advance). In 1950, John Nash extended the possibilities to non-zero-sum games by introducing the mathematical description of a general solution for non-cooperative games, later called the Nash equilibrium [160]. With these new possibilities, game theory became a dominant modeling method for decision making in economics and later in social sciences. In 1994, Nash was awarded the Nobel Prize together with Reinhard Selten and John Harsanyi for his findings in game theory. Several Nobel Prize awards are expected to follow for important enhancements in game theory approaches.

Over time the assumption of the "homo oeconomicus", a fully rationally acting human being, became questioned. Herbert Simon's research work showed that not only rationality, but also criteria like envy, greed or fairness can be part of the human decision [161]. In 1978, Simon was awarded the Nobel Prize in Economics for his findings.

In 2002, Daniel Kahneman was awarded the same prize for his experimentally based theory about spontaneous and situation-based decision making. According to this theory, humans often act reciprocally, which can result in a better outcome than what the "homo oeconomicus" could reach by strictly rational behavior. In addition, Kahneman showed that mutually cooperative behavioral strategies occur and can prevail [162]. Robert Axelrod developed ideas about the evolution of cooperation and e.g. showed how they apply to warfare [163]. Two more Nobel Prizes for works in the field of game theory were awarded to Thomas Schelling and Robert Aumann in 2005.

Biological use cases represented a significant enhancement to game theory applications. Evolutionary game theory states that human behavior is also resulting from genetically and culturally influenced processes, and not by pure rationality only.

Game theory allows modeling and mathematically solving strategic games and situations of conflict, which typically represent complex systems. A major point of criticism on game theory is that partly unrealistic assumptions need to be made. Especially the rational human behavior in complex situations has been shown as being unrealistic in recent times. Different further developed approaches of game theory therefore try to overcome such shortcomings.

4.4   Discussion of Historical Developments

Nowadays, managing complexity possesses a significant importance for many people. Compared to former times, job-related as well as private situations can be characterized by complexity. But already more than 2000 years ago, situations occurred that asked for managing complexity. In fact, the challenge of solving complexity can be seen during all periods of science history. For that reason it is not surprising that also at all times, procedures have been created for complexity management.

There are differing opinions on when complexity became such a relevant issue that a means for its management has been actively developed. And several major historical events meant a significant increase of complexity, for example the beginning of intercontinental trading, the industrial revolution or the two world wars. Another boost of complexity seemed just to have happened with the quick spreading of the Internet and interconnectivity by popular social networks.

The uprising of systems thinking in the early twentieth century stands out in the historic development of complexity management, as it prepared the basis for a systematic and scientific approach. Developing scientific methods and procedures was largely motivated and facilitated by the necessities of the Second World War. Those findings and experiences generated from war times then built the fundaments of modern approaches towards complexity management, which is also applied to civil applications.

For many findings, developments and innovations, war situations acted as the catalyzer and boosted scientific progress. This also accounts for system sciences and complexity management. And many developments in this discipline can even be traced back to ancient Greek philosophy. So, the historical developments described in this chapter indicate that modern methods of complexity management are the result of an evolutionary, not revolutionary process.

As the basics of modern complexity management have already been created in the 1940s, why did it take complexity management until recent years to reach public visibility? Two reasons can be found for this.

Obviously and already mentioned above, complexity increased dramatically in everybody's life in the last decades. Complexity did not only increase for governments, large organizations and enterprises, but also individuals experienced complexity increases in their jobs and private lives—mainly driven by an increase of information to be processed. This increased complexity naturally raised the demand for applicable methods for complexity avoidance, reduction and management.

The other reason is that scientists researching on systems and complexity in the middle of the twentieth century were way ahead of their time. It took time for technical equipment like computers to be developed and get distributed in large numbers. And as today every PC brings software for modeling a system network, this took half a century to get to this point. The same can be said about the use cases associated with complexity. Today, effects of highly interlinked networks are a common experience in everybody's life, for example when news, videos or pictures go viral on social networks. Fifty years ago, the notion of thinking in networks was rather abstract and was only a part of few people's daily lives.

In many ways, the breakthrough of complexity management is closely related to the development of the computer. While an increase of complexity can be associated with the triumphant success of computers, they also represent a fundamental prerequisite for solving complex challenges. Therefore it is not surprising that system scientists like John von Neumann and Heinz von Foerster were also deeply involved in the development of the computer. And only today's worldwide availability of computers allows many people to address complex challenges themselves.

References

1. Schwanitz, D. 1999. *Bildung. Alles, Was Man Wisen Muss*. Frankfurt: Eichborn.
2. Wiener, Norbert. 1948. *Cybernetics or Control and Communication in the Animal and the Machine*. New York: Technology Press.
3. Husbands, Phil, and Owen Holland. 2008. The Ratio Club: A Hub of British Cybernetics. In *The Mechanical Mind in History*, ed. P. Husbands, O. Holland, and M. Wheeler, 91–148. Cambridge, MA: MIT Press.
4. Laszlo, E. 1972. *Introduction to Systems Philosophy—Toward a New Paradigm of Contemporary Thought*. New York: Gordon and Breach.
5. Bertalanffy, Ludwig v. 1975. *Perspectives on General System Theory—Scientific-Philosophical Studies*. New York: George Braziller.
6. Johannessen, S.O., and L. Kuhn, eds. 2012. *Complexity in Organization Studies*. Vol. 1. Los Angeles: Sage.
7. Capra, F. 1984. *Wendezeit. Bausteine Für Ein Neues Weltbild*. 6th ed. Bern: Scherz.
8. Schalcher, H.R. 2008. Systems Engineering. Vorlesungsunterlagen. http://www0.ibb.ethz.ch/de/lehre/pm_unterlagen/SE/SE-Skript0809.pdf.
9. Plato. 2006. *Timaeus*. Edited and translated by B. Jowett. Teddington: The Echo Library.
10. Smuts, J.C., and A. Meyer-Abich. 1938. *Die Holistische Welt*. Berlin: Alfred Metzner.
11. Audi, Robert. 1999. *The Cambridge Dictionary of Philosophy*. 2nd ed. Cambridge, UK: Cambridge University Press.
12. Nagel, Ernest. 1960. Determinism in History. *Philosophy and Phenomenological Research* 20 (3): 291–317.
13. Chalmers, Alan. 2014. Atomism from the 17th to the 20th Century. Stanford Encyclopedia of Philosophy. http://plato.stanford.edu/entries/atomism-modern/.
14. Mintz, S.J. 2010. *The Hunting of Leviathan: Seventeenth-Century Reaction of the Materialism and Moral Philosophy of Thomas Hobbes*. Cambridge: Cambridge University Press.
15. Copley, F.B. 2010. Frederick W. Taylor—Father of Scientific Management. Bertrams Print on Demand.
16. Cunningham, Andrew, and Nicholas Jardine. 1990. Romanticism and the Sciences. CUP Archive.
17. Driesch, Hans. 1905. *Der Vitalismus Als Geschichte Und Als Lehre*. Leipzig: Verlag von Johann Ambrosius Barth.
18. Cimino, G., and F. Ducnesneau, eds. 1993. Vitalisms from Haller to the Cell Theory. In Proceedings of the Zaragoza Symposium XIXth International Congress of History Os Science. Firenze: Olschki, Leo S.
19. Diersch, H. 2009. Der Vialismus Als Geschichte Und Lehre. Biblio Bazaar.

20. Bertalanffy, Ludwig v. 1950. An Outline of General Systems Theory. *The British Journal for the Philosophy of Science* 1/2: 134–165.
21. ———. 1973. *General System Theory: Foundations, Development, Applications.* 4th ed. - New York: George Braziller.
22. Ramage, M., and K. Ship. 2009. *System Thinkers.* Dordrecht: Springer.
23. Bertalanffy, Ludwig v. 1934. *Modern Theories of Development.* Edited and translated by J. H. Woodger. Oxford: Oxford University Press.
24. ———. 1972. The History and Status of General Systems Theory. *The Academy of Management Journal* 15(4): 407–426.
25. Matthies, M., and J. Zimmermann. 2010. Für Eine Geschichte Der Systemwissenschaft. Universität Osnabrück. http://www.usf.uos.de/usf/literatur/beitraege/texte/053-zimmermann. pdf.
26. Drack, M. 2008. Ludwig von Bertalanffy's Early System Approach. Universität Wien.
27. Luhmann, N. 2012. *Introduction to Systems Theory.* Cambridge: Wiley.
28. Churchill, W.S. 1948. *The Second World War.* Reissue Ed. Boston, MA: Mariner Books.
29. Galison, P. 2001. Die Ontologie Des Feindes: Norbert Wiener Und Die Kybernetik. In *Ansichten Der Wissenschaftsgeschichte,* ed. M. Hagner. Frankfurt a. M.: Fischer Taschenbuch.
30. Bluma, L. 2004. *Norbert Wiener Und Die Entstehung Der Kybernetik Im Zweiten Weltkrieg.* Münster: LIT.
31. Pruckner, M. 2002. 90 Jahre Heinz von Förster. Die Praktische Bedeutung Seiner Wichtigsten Arbeiten. Malik Management Zentrum St. Gallen.
32. Time Magazine Cover Mark III. 1950. http://content.time.com/time/covers/0,16641,19500123,00. html.
33. Young, J.F. 1973. *Cybernetic Engineering.* London: Butterworths.
34. Asc-cybernetics.org. 2015. *Summary: The Macy Conferences at Asc-Cybernetics.org.* Accessed 20 Dec. http://www.asc-cybernetics.org/foundations/history/MacySummary.htm.
35. von Foerster, Heinz, Margaret Mead, and Hans Lukas Teuber. 1950. No Title. In *Cybernetics: Transactions of the Seventh Conference.* New York: Josiah Macy, Jr. Foundation.
36. ———. 1952. No Title. In *Cybernetics: Transactions of the Eighth Conference.* New York: Josiah Macy, Jr. Foundation.
37. ———. 1953. No Title. In *Cybernetics: Transactions of the Ninth Conference.* New York: Josiah Macy, Jr. Foundation.
38. ———. 1955. No Title. In *Cybernetics: Transactions of the Tenth Conference.* New York: Josiah Macy, Jr. Foundation.
39. Dupuy, Jean-Pierre. 2000. *The Mechanization of the Mind: On the Origins of Cognitive Science.* Edited and translated by M.B. DeBevoise. Princeton, NJ: Princeton University Press.
40. Heims, Steve J. 1991. *The Cybernetics Group.* Cambridge, MA: MIT Press.
41. von Foerster, Heinz. 1960. On Self-Organizing Systems and Their Environments. In *Self-Organizing Systems,* 31–50. London: Pergamon Press.
42. Müller, A. 2000. Eine kurze Geschichte des BCL—Heinz von Förster und das Biological Computer Laboratory. *Österreichische Zeitschrift Für Geschichtswissenschaften* 11(1): 9–30.
43. Müller, A., and K. Müller, eds. 2007. *An Unfinished Revolution? Heinz von Foerster and the Biological Computer Laboratory (BCL), 1958–1976.* Vienna: Edition Echoraum.
44. von Foerster, Heinz, and George W. Jr. Zopf. 1962. No Title. In *Principles of Self-Organization. Transactions of the University of Illinois Symposium on Self-Organization.* New York: Symposium Publications Pergamon Press.
45. Pickering, A. 2004. The Science of the Unknowable: Stafford Beer's Cybernetic Informations. *Kybernetes* 33(3/4): 499–521.
46. Beer, Stafford. 1995. *Brain of the Firm.* 2nd ed. New York: Wiley.

47. Vester, Frederic. 2007. *The Art of Interconnected Thinking—Ideas and Tools for Tackling Complexity*. Munich: Mcb Verlag.
48. ———. 2015. Sensitivity Model. Accessed 22 Dec. http://www.frederic-vester.de/eng/sensitiv ity-model/.
49. Jackson, Michael C. 2000. *Systems Approaches to Management*. Berlin: Springer.
50. Rosenhead, J. 2006. IFORS Operational Research Hall of Fame Stafford Beer. *International Transactions in Operational Research* 13(6): 577–581.
51. Beer, Stafford. 1959. *Cybernetics and Management*. New York: Wiley.
52. ———. 1994. *Decision and Control—The Meaning of Operational Research and Management Cybernetics*. New York: Wiley.
53. ———. 1972. *The Brain of the Firm*. London: Allen Lane.
54. ———. 1979. *The Heart of Enterprise*. New York: Wiley.
55. ———. 1984. The Viable System Model: "Its Provenance, Development, Methodology and Pathology". *The Journal of Operational Research Society* 35(1): 7–25.
56. Bröker, J. 2005. Erfolgreiches Management Komplexer Franchisesysteme Auf Grundlage Des Viable System Model. Universität St. Gallen.
57. Jackson, Michael C. 1988. An Appreciation of Stafford Beer's "Viable System" Viewpoint on Managerial Practice. *Journal of Management Studies* 25(6): 557–573.
58. Beer, Stafford. 1994. *Beyond Dispute. The Invention of Team Syntegrity*. New York: Wiley.
59. Malik, F. 2006. *Führen Leisten Leben—Wirksames Management Für Eine Neue Zeit*. Frankfurt a Main: Campus.
60. Ulrich, H., and G. Probst. 1984. *Self-Organization and Management of Social Systems: Insights, Promises, Doubts, and Questions*. Berlin: Springer.
61. Probst, G., and P. Gomez. 1992. Thinking in Networks to Avoid Pitfalls of Management Thinking. In *Context and Complexity*, 91–108. New York: Springer.
62. Riggs, J.L., and M.S. Inoue. 1975. *Introduction to Operations Research and Management Science: A General Systems Approach*. New York: McGraw-Hill.
63. Churchman, C.W., R.L. Ackoff, and E.L. Arnoff. 1959. *Introduction in Operations Research*. 4th ed. New York: Wiley.
64. Kirby, M.W. 2003. *Operational Research in War and Peace*. London: World Scientific.
65. Shrader, C.R. 2006. History of Operations Research in the United States Army, Vol. I: 1942–1962. Washington, DC: United States Army. http://www.history.army.mil/html/books/hist_op_research/CMH_70-102-1.pdf.
66. Schlee, W. 2001. *Unternehmensforschung*. München: Technische Universität München, Zentrum Mathematik.
67. Rajgopal, J. 2004. Principles and Applications of Operations Research. In *Industrial Engineering Handbook*, 5th ed., 11.27–11.44. New York: McGraw-Hill. http://www.pitt.edu/~jrclass/or/or-intro.html.
68. Hahn, F. 2006. *Von Unsinn Bis Untergang: Rezeption des Club of Rome und Grenzen des Wachstums in der Bundesrepublik der frühen 1970er Jahre*. Freiburg: Albert-Ludwigs-Universität.
69. Gass, S.I. 1985. *Linear Programming, Methods and Applications*. 5th ed. New York: McGraw-Hill.
70. Zimmermann, W., and U. Stache. 2001. *Operations Research—Quantitative Methoden Zur Entscheidungsvorbereitung*. 10th ed. München: Oldenbourg Wissenschaftsverlag.
71. RAND Corporation. 2015. http://www.rand.org/. Accessed 29 Dec.
72. Gass, S.I., and C.M. Harris, eds. 2001. *Encyclopedia of Operations Research and Management Silence*. 2nd ed. New York: Springer.

73. Marvel, O.E. 2007. The History of System Engineering Methodologies. Naval Post Graduate School Monterey.
74. Brill, James H. 1998. Systems Engineering? A Retrospective View. *Systems Engineering* 1(4): 258–266. doi:10.1002/(SICI)1520-6858(1998)1:4<258::AID-SYS2>3.0.CO;2-E.
75. Hall, A.D. 1962. *A Methodology for Systems Engineering*. Princeton, NJ: Nostand.
76. Goode, Harry, and Robert Machol. 1957. *System Engineering: An Introduction to the Design of Large-Scale Systems*. New York: McGraw-Hill.
77. Hall, A.D. 1989. *Metasystems Methodology*. New York: Pergamon.
78. Chestnut, H. 1965. *Systems Engineering Tools*. New York: Wiley.
79. Shinners, S. 1967. *Techniques of Systems Engineering*. New York: McGraw-Hill.
80. Miles, R.F. 1973. *Systems Concepts*. New York: Wiley.
81. Chase, W.P. 1974. *Management of Systems Engineering*. Malabar, FL: Robert Krieger.
82. Wymore, W. 1976. *Systems Engineering Methodology for Interdisciplinary Teams*. New York: Wiley.
83. Sage, A.P. 1995. *Systems Management for Information Technology and Software Engineering*. New York: Wiley.
84. Sage, Andrew P. 1977. *Methodology for Large-Scale Systems*. New York: McGraw-Hill.
85. Blanchard, B., and W. Fabrycky. 1981. *Systems Engineering and Analysis*. Englewood Cliffs, NJ: Prentice-Hall.
86. Booton, Richard C. Jr., and S. Ramo. 1984. The Development of Systems Engineering. *IEEE Transactions AES* 20–24(July): 306–311.
87. Welgel, Annalisa. 2000. An Overview of the Systems Engineering Knowledge Domain. MIT. http://web.mit.edu/esd.83/www/notebook/sysengkd.pdf
88. Air Force Systems Command Headquarters. 1966. *AFSC Manual 375-5, Systems Engineering Management Procedures*. Washington, DC: Andrews Air Force Base.
89. National Astronautics and Space Administration (NASA). 1995. *Systems Engineering Handbook*. Washington, DC.
90. Headquarters, and Department of the Army. 1979. Field Manual 770-78, Systems Engineering. Washington, DC.
91. MIL-STD-499 (USAF). 1969. Systems Engineering Management. Washington, DC: Department of Defense.
92. MIL-STD-499A (USAF). 1974. Engineering Management. Washington, DC: Department of Defense.
93. MIL-STD-499B (USAF). 1992. Systems Engineering. For Coordi. Washington, DC: Department of Defense.
94. Systems Engineering Management Guide. 1983. Ft. Belvoir, VA: Defense Systems Management College (DSMC).
95. Electronic Industries Association (EIA). 1994. *Interim Standard 632*. Washington, DC: Systems Engineering.
96. Engineers, Institute of Electrical and Electronics. 1993. IEEE, P1220, Standard for Systems Engineering. New York.
97. IEEE, 1220-1994, Trial-Use Standard for Application and Management of the Systems Engineering Process. 1995. New York: Institute of Electrical and Electronics Engineers.
98. INCOSE. 2015. *INCOSE Systems Engineering Handbook, A Guide for System Life Cycle Processes and Activities*. 4th ed. Hoboken, NJ: Wiley.
99. Gorod, A., B. Sauser, and J. Boardman. 2008. System-of-Systems Engineering Management: A Review of Modern History and a Path Forward. *IEEE Systems Journal* 2(4): 484–499. doi:10. 1109/JSYST.2008.2007163.

100. Keating, C., R. Rogers, R. Unal, D. Dryer, A. Sousa-Poza, R. Safford, W. Peterson, and G. Rabaldi. 2003. Systems of Systems Engineering. *Engineering Management Journal* 15(3): 36–45.

101. Eisner, H., J. Marciniak, and R. McMillan. 1991. Computer-Aided System of Systems (C2) Engineering. In IEEE International Conference on Systems Man and Cybernetics. Charlottesville, VA.

102. Eisner, H. 1993. RCASSE: Rapid Computer-Aided Systems of Systems (S2) Engineering. In 3rd International Symposium of the National Council on System Engineering, 267–273.

103. Shenhar, A. 1994. A New Systems Engineering Taxonomy. In 4th International Symposium of the National Council on System Engineering, 261–276.

104. Holland, John H. 1995. *No TitleHidden Order: How Adaptation Builds Complexity*. Reading, MA: Addison-Wesley.

105. Owens, A.W.A. 1995. The Emerging U.S. System of Systems. In *Dominant Battlespace Knowledge*, ed. S. Johnson and M. Libicki. Washington, DC: NDU Press.

106. Manthorpe, W.H. 1996. *The Emerging Joint System of System: A Systems Engineering Challenge and Opportunity for APL*. John Hopkins APL Technical Digest, vol. 17, 305–310.

107. Maier, M. 1996. Architecting Principles of Systems-of-Systems. In 6th Annual International Symposium of the International Council on System Engineering. Boston, MA.

108. Maier, Mark W. 1998. Architecting Principles for Systems-of-Systems. *Systems Engineering* 1 (4): 267–284. doi:10.1002/(SICI)1520-6858(1998)1:4<267::AID-SYS3>3.0.CO;2-D.

109. Kotov, V. 1997. Systems of Systems as Communicating Structures. HPL-97th–124th ed. Hewlett Packard.

110. Luskasik, S.J. 1998. System, Systems of Systems, and the Education of Engineers. *Artificial Intelligence for Engineering Design, Analysis and Manufacturing* 12(1): 55–60.

111. Pei, R. 2000. System of Systems Integration (SoSI)—A Smart Way of Acquiring Army C412WS Systems. In Summer Computer Simulation Conference, 574–579.

112. Carlock, Paul G., and Robert E. Fenton. 2001. System of Systems (SoS) Enterprise Systems Engineering for Information-Intensive Organizations. *Systems Engineering* 4(4): 242–261. doi:10.1002/sys.1021.

113. Cook, S.C. 2001. On the Acquisition of Systems of Systems. In INCOSE Annual Symposium, Melbourne.

114. Sage, A.P., and C.D. Cuppan. 2001. On the Systems Engineering and Management of Systems of Systems and Federations of Systems. *Information Knowledge System Management* 2: 325–345.

115. Chen, Pin, and Jennie Clothier. 2003. Advancing Systems Engineering for Systems-of-Systems Challenges. *Systems Engineering* 6(3): 170–183. doi:10.1002/sys.10042.

116. Bar-Yam, Y., M.A. Allison, R. Batdorf, H. Chen, H. Generazio, H. Singh, and S. Tucker. 2004. The Characteristics and Emerging Behaviors System of Systems. In *NECSI: Complex Physical, Biological and Social Systems Project*.

117. Jamshidi, M. 2005. System of Systems engineering—A Definition. In IEEE System, Man and Cybernetics. Piscataway, NJ.

118. Lane, J.A., and R. Valerdi. 2005. Synthesizing SoS Concepts for Use in Cost Estimation. In IEEE Conference on System, Man and Cybernetics. Waikoloa, HI.

119. Boardman, J., and B. Sauser. 2006. System of systems—The Meaning of oF. In IEEE International Systems of Systems Engineering Conference. Los Angeles, CA.

120. Jamshidi, M., eds. 2008. *System of System Engineering—Innovations for the 21st Century*. Hoboken, NJ: Wiley.

121. ———., eds. 2008. *System of Systems—Principles and Applications*. Boca Raton, FL: Taylor & Francis.

122. Johnson, S.B. 2006. *The Secret of Apollo: System Management in American and European Space Programs*. Baltimore, MD: The John Hopkins University Press.
123. AIAA Systems Engineering Technical Committee. 2015. Accessed 29 Dec. https://info.aiaa.org/tac/ETMG/SETC/default.aspx.
124. IEEE Systems Council. 2015. http://ieeesystemscouncil.org/. Accessed 29 Dec.
125. Walter, U. 2012. Systems Engineering Vorlesung. Technische Universität München.
126. Blanchard, B.S. 2008. *System Engineering Management*. 4th ed. Hoboken: Wiley.
127. Zachman International. 2016. https://www.zachman.com/about-the-zachman-framework.
128. Zachman, John. 1987. A Framework for Information Systems Architecture. *IBM Systems Journal* 26(3): 454–470.
129. The Open Group—TOGAF. 2016. http://www.opengroup.org/subjectareas/enterprise/togaf. Accessed 5 July.
130. DoD Architecture Framework Version 2.02. 2011. http://dodcio.defense.gov/Portals/0/Documents/DODAF/DoDAF_v2-02_web.pdf.
131. AUTOSAR. 2016. http://www.autosar.org. Accessed 13 July.
132. Schekkerman, Jaap. 2003. *How to Survive in the Jungle of Enterprise Architecture Frameworks: Creating or Choosing an Enterprise Architecture Framework*. Bloomington, IN: Trafford Publishing.
133. ISO/IEC/IEEE 42010:2011, Systems and Software Engineering—Architecture Description. 2011. http://www.iso-architecture.org/42010/index.html.
134. Forrester, Jay W. 1989. The Beginning of System Dynamics. https://www.google.com/url?sa=t&rct=j&q=&esrc=s&source=web&cd=1&ved=0ahUKEwj_-t3h8YTKAhUT_cz4KHZ-hBJUOFggfMAA&url=http%3A%2F%2Fweb mit edu%2Fsysdyn%2Fsd-intro%2FD-4165-1 pdf&usg=AFQjCNGHOrDJYvx3K-F0au7cfycx0vOZnA&cad=rjt.
135. ———. 1992. From the Ranch to System Dynamics: An Autobiography. In *Management Laureates—A Collection of Autobiographical Essays*, ed. Arthur G. Bedeian, vol. 1. Greenwich, CT: JAI Press.
136. Elichirigoity, F.J. 1999. *Planet Management: Limits of Growth, Computer Simulation, and the Emergence of Global Spaces*. Evanston: Northwestern University Press.
137. Kapmeier, F. 1999. Vom Systemischen Denken Zur Methode System Dynamics. Universität Stuttgart.
138. Forrester, Jay W. 1958. Industrial Dynamics: A Major Breakthrough for Decision Makers. *Harvard Business Review* 36(4): 37–66.
139. ———. 1971. *World Dynamics*. Cambridge: Wright-Allen Press.
140. Coyle, R.G. 1996. *System Dynamics Modelling—A Practical Approach*. London: Chapman & Hall.
141. ———. 2000. Quantitative and Qualitative Modeling in System Dynamics: Some Research Questions. *System Dynamic Review* 16(3): 225–244.
142. Sterman, John D 2000. On an Approach to Techniques for the Analysis of the Structure of Large Systems of Equations. *SIAM Review* 4(4).
143. Forrester, Jay W. 1969. *Urban Dynamics*. Cambridge, MA: MIT Press.
144. Deutsche Gesellschaft Für System Dynamics e.V. 2015. http://www.systemdynamics.de/. Accessed 29 Dec.
145. System Dynamics Society. 2015. http://www.systemdynamics.org/. Accessed 29 Dec.
146. Meadows, D.H. 1972. *The Limits to Growth*. Reissue. Verlag Signet.
147. Rüffer, Th. 2012. *Der System-Dynamics-Ansatz als Untersuchungsverfahren im Rahmen des Managements eines Projektes*. Berlin: Springer.
148. Schwarz, R., and J.W. Ewaldt. 2005. *Über Den Beitrag Systemdynamischer Modellierung Zur Abschätzung Technologischer Evolution*. Berlin: Springer.

149. Wheat, J.D. Jr. 2007. *The Feedback Method—A System Dynamics Approach to Teaching Macroeconomics*. Norway: University of Bergen.
150. Gabler Wirtschaftslexikon. 2015. http://wirtschaftslexikon.gabler.de/Archiv/143837/system-dynamics-v5.html. Accessed 29 Dec.
151. Thiel, Daniel. 1994. System Dynamics in Educational Science: An Experience of Teaching Production-Distribution Mental Models Building. In 1994 International System Dynamics Conference, 95–104. http://www.systemdynamics.org/conferences/1994/proceed/papers_vol_2/thiel095.pdf.
152. Deco, G., and B. Schürmann. 2001. *Information Dynamics: Foundations and Applications*. New York: Springer.
153. Sterman, John D. 2015. Teaching Takes Off: Flight Simulators for Management Education—'The Beer Game.' http://web.mit.edu/jsterman/www/SDG/beergame.html. Accessed 29 Dec.
154. ———. 1989. Modeling Managerial Behavior: Misperceptions of Feedback in a Dynamic Decision Making Experiment. *Management Science* 35(3): 321–339.
155. Osborne, M.J. 2003. *An Introduction to Game Theory*. 7th ed. Oxford: Oxford University Press.
156. Avenhaus, R., F. Lehmann, and A. Wölling. 1997. Anwendungen Der Spieltheorie. Studentenprotokoll Zum Oberseminar Im Anwendungsfach Informatik. Universität der Bundeswehr München.
157. von Neumann, John, and Oskar Morgenstern. 1944. *Theory of Games and Economic Behavior*. Princeton, NJ: Princeton University Press.
158. Borel, Émile. 1921. La Théorie Du Jeu et Les Équations Intégrales À Noyau Symétrique. *Comptes Rendus Hebdomadaires Des Séances de l'Académie Des Sciences* 173(7): 1304–1308.
159. Poundstone, William. 1992. *Prisoner's Dilemma: John Von Neumann, Game Theory and the Puzzle of the Bomb*. New York: Doubleday.
160. Nash, J. 1950. Non-Cooperative Games. Princeton, NJ: Princeton University Press. www.princeton.edu/mudd/news/faq/topics/Non-Cooperative_Games_Nash.pdf.
161. Simon, Herbert Alexander. 1992. Economics, Bounded Rationality and the Cognitive Revolution. Edited by M. Egidi and R.L. Marris. Cheltenham: Edward Elgar.
162. Kahnemann, D. 2011. *Thinking, Fast and Slow*. London: Macmillan.
163. Axelrod, R.M. 1984. *The Evolution of Cooperation*. New York: Basic Books.

Classification of Complexity Management Approaches in Engineering

Earlier sections of this thesis pointed out that complexity challenges appear in many different fields and are tackled by a variety of strategies and methods. The lack of knowledge, which is often associated with complex problems, has been addressed in literature by several authors. Craig Read claims that industry has no understanding of complexity. Neither the origins nor the effects of it are addressed, and approaches to complexity are missing [1]. Sheard and Mostashari urge that future work should fill in gaps in understanding complexity, especially if related to systems engineering [2].

As systems engineering is about realizing successful systems, dealing with complexity seems to be inevitable. In fact, in many publications in this field the authors complain about the increasing complexity. This led to the proposition of a new sub-discipline called "complex systems engineering" [3].

Whereas in engineering the relevance of complex system interactions is obvious, the meaning of complexity and complexity management is still indistinct. In order to provide transparency over existing approaches and methods, this chapter presents a map of fields in systems engineering and complexity management topics and approaches within these fields. The map further depicts overlaps between the fields based on similarity in complexity management practices. The sheer amount of publications makes it impossible to depict all trends and developments. Therefore, this chapter describes selected works of authors who significantly contributed to and influenced their scientific fields. This shall illustrate the evolution of different research fields and their interconnectivity regarding complexity. And it shall facilitate the transfer of methods and procedures of complexity management, as the application of new methods is often inspired by transferring them from other fields.

Because of the high popularity of the term complexity in a multitude of research fields, the search was restricted to publications with technical, engineering or systems engineering background. Initially a variety of terms connected to the topic of complexity, e.g. complex systems, complexity management or complex engineering were identified. Then relevant

© Springer-Verlag GmbH Germany 2017
M. Maurer, *Complexity Management in Engineering Design – a Primer*,
DOI 10.1007/978-3-662-53448-9_5

topics and authors were identified and grouped, and dependencies between topics were highlighted as overlaps. Such an overlap mean that an identified publication belongs to at least two different main topics. Observing the overview map does not only show overlaps between topics—also missing overlaps can be of interest, as such "blind spots" can represent important future research topics.

5.1   A Map of Complexity Management Approaches

By a literature review, seven disciplines could be identified in the engineering field which have a direct relevance to issues of complexity.

Figure 5.1 shows these seven disciplines as circles in a Venn diagram. "Complex systems" is depicted in the center of the diagram and by a larger circle than the other six disciplines. This is a result of the many overlaps of this discipline with all the other ones. It should not intend to convey a higher importance of this discipline. Section 5.2 describes the seven disciplines, key authors and their complexity management approaches.

Besides the main disciplines, the research unveiled overlaps between the various disciplines and in turn helped to expose voids or blind spots between the disciplines. Eleven overlaps between disciplines are depicted in Fig. 5.1 and indicated with Roman numerals. Nine overlaps result from the interaction between two disciplines, and two

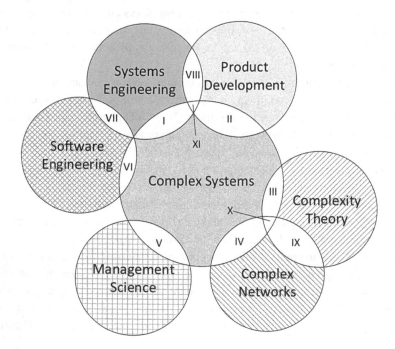

Fig. 5.1 Complexity overview Venn diagram

overlaps from even three disciplines. Section 5.3 describes the overlaps in detail, referring to the numerals of the diagram. And Sect. 5.4 addresses blind spots on this map and discusses the findings.

It must be mentioned that a diagram, especially a simple Venn diagram, can hardly cope with the many aspects of complexity. That means that Fig. 5.1 can only depict a partial view. As well, the enormous amount of publications dealing with many different aspects and perspectives of complexity cannot be comprehensively visualized in one picture. Consequentially, Fig. 5.1 must be incomplete. And depending on one's personal perspective, different classifications can be more relevant. Nevertheless, the chosen visualization of complexity disciplines can be helpful for understanding this highly important topic. The figure facilitates the interpretation and discussion of complexity aspects. By providing a picture that is easy to understand, it may provoke the reader to challenge, adapt and extend it. So this chapter and the included descriptions and explications should be seen as a work in progress, with the objective of better understanding complexity management in engineering.

5.2   The Main Research Areas of Engineering Complexity

5.2.1   Product Development

The complex nature of today's engineering systems often brings forth the issue of complexity in product development. Eppinger states that "Development of complex products and large systems is a highly interactive social process involving hundreds of people designing thousands of interrelated components and making millions of coupled decisions" [4]. He introduces "three views of product development complexity: a process view, a product view, and an organization view", and explains that one can "learn about the complex social phenomenon of product development by studying the patterns of interaction across the decomposed elements within each view". That means that Eppinger considers structural complexity that emerges from large numbers of dependencies between system elements, which results in hardly predictable interactions within those complex (sub-)systems.

Focusing on the structure of the product, process and organizational views allows for decomposing "in order to manage the complexity". Thus, decomposing the product into several subsystems makes its complexity more understandable as smaller problems can be solved more easily, hence speeding up the problem-solving and solution identification process. According to Eppinger, decomposing and analyzing these three system views can lead to the following benefits when designing complex systems [4]:

The product view comprises dependencies between technical components. A detailed analysis of these dependencies can point to more effective module and sub-system boundaries, which can make the whole system easier to handle. Furthermore, the analysis can support interface management, as critical interfaces can be discovered. And as a result of well-known module boundaries, appropriate opportunities for outsourcing can be identified.

The investigation of the process view can be helpful for streamlining process steps, which then can lead to reduced process times. Design iterations can also be identified by analyzing the process view. And if these iterations can be avoided or at least reduced, this can further accelerate the design process. Eppinger further mentions that failure modes within the process can be seen from the process view and that chaotic information flows, when identified, can be replaced by more effective formal procedures.

Finally, the decomposition and analysis of the organizational view can allow for more effective team arrangements. And system engineering functions can be applied for better integration of the overall product [4].

Not only the consideration of the three isolated system views can help manage the system complexity. Also analyzing the comparison views between two system views can "serve to help diagnose cultural and dynamic causes of process-related and organizational failures to efficiently develop the selected product architecture" [4]. The complexity of the process is directly proportional to the complexity of the product architecture; therefore, more complex architectures require more complex processes for their development. The organizational structure domain synchronizes the product development activities; hence, the organizational structure itself is split into different organizational groups with different skills. Eppinger's considerations are in line with an earlier finding known as "Conway's law", stating that the product design structures result from the organizational structures because of the interactions by communication [5].

Ulrich focuses on product architecture development and states that unlike modular product architecture "an integral architecture includes a complex (non one-to-one) mapping from functional elements to physical components and/or coupled interfaces between components" [6]. He further mentions that component standardization represents a possibility for reducing complexity, but this has to be balanced with potential deficits e.g. associated with product performance and costs. Ulrich links the skills required in an organization structure to the product architecture and emphasizes that "highly modular designs allow firms to divide their development and production organizations into specialized groups with a narrow focus" and thus require better systems engineering and planning skills, whereas integral architectures require better coordination and integration skills [6].

Like Eppinger, also Ulrich investigates issues of structural complexity, in his case mostly the product architecture view, which is linked to other system parts like process and organizational view. Because of these links, constellations or measures in the product architecture can e.g. result in an increase in management complexity.

Structural complexity results from system elements, e.g. components, process steps or organizational units, which are interrelated by e.g. communication flows, change propagation or material flow. Dependency modeling (see also Sect. 3.4.4) provides several methods and tools, which have been successfully applied in numerous projects. Besides well-established visualization and computation approaches for process and organizational structures (e.g. flowcharts or event-driven process chains), the Design Structure Matrix (DSM) has especially gained popularity as a generic method for system structure modeling and optimization [7]. An extension of the DSM, called Multiple Domain Matrix (MDM), allows for integrated consideration of several product views, as it has been proposed by several authors, [4, 8, 9]. Lindemann et al. introduced an approach using an MDM and guiding the user from acquiring structural information until the identification of structural improvements—and called it "Structural Complexity Management" [10].

Complexity management approaches in product development consider different aspects of the product generation and mostly focus on structural complexity. Many methods and tools for its visualization, analysis and optimization get applied. And many approaches, e.g. modular design and interface management comprise those methods.

5.2.2   Systems Engineering

Systems engineering is an interdisciplinary field that emerged to manage complex engineering systems, hence it has strong link to the field of complex systems. From a historical point of view, the Guide to the Systems Engineering Body of Knowledge states: "We can view the evolution of systems engineering (SE) in terms of challenges and responses. Humans have faced increasingly complex challenges and have had to think systematically and holistically in order to produce successful responses to challenges. From these responses, generalists have developed generic principles and practices for replicating success" [11].

Several definitions of systems engineering are available, e.g. by [12, 13], and complexity is often mentioned in this context. For a structured discussion about the management of complexity in systems engineering, Sheard states "that what people have been calling 'systems engineering' can be split into three basic implementations or types of systems engineering: Discovery, a discipline or specialist type that involves significant analysis, particularly of the problem space; Program Systems Engineering, a coordination or generalist type that emphasizes the solution space and technical and human interfaces; and Approach, a process type that can (and should) be performed by any engineer" [14]. In these three types of systems engineering, complexity management play different roles and is realized with different approaches.

"The models and analyses created in Discovery implementations include requirements modeling, system behavior modeling, environmental and system simulation, reliability or survivability analysis, orbital analyses (for space systems), and contingency scenarios, to name a few" [14]. Many modeling approaches also known in other areas get applied, and methods for managing structural complexity are of significant importance. Behavioral modeling and simulations largely deal with dynamic complexity and apply approaches from the field of system dynamics.

In Program Systems Engineering "the problem space is more precedented (commercial communication satellites, for example), and the unprecedented aspect is how the pieces fit together to provide a new variation on a type of service." "The emphasis in Program Systems Engineering is on producing cost-effective solutions that meet quality and schedule criteria. Organizational processes can be defined and improved to standardize this type of systems engineering" [14]. This type of systems engineering has to deal with architectural, structural and interface questions and is very close to the complexity-managing approaches of product development. Therefore methods for modeling, analyzing and optimizing system structures can get applied successfully.

"Approach is the systems engineering that every engineer must perform on the product, including understanding risks, understanding the operational need, clarifying requirements before jumping to a solution, doing at least informal trade studies to make decisions, and so on through the process or focus areas of the chosen capability model" [14]. For this systems engineering approach, complexity challenges apply to the actions in the responsibility of an engineer. Thus, methods and tools that create transparency and help understanding the impact and consequences of decisions can be applied.

Abbott states that systems engineering is the engineering of complex systems as a result of the multi-scale interaction among the system components due to the integration of hardware, software and services [15]. The systems are designed for uncertainties and include adaptation strategies that enhance reliability and robustness.

Sheard stresses that the traditional trend of systems engineering is metamorphosing into complex systems engineering, and suggests that systems engineering bodies such as INCOSE can contribute by modifying themselves to enable engineers to use complex systems ideas [16].

5.2.3 Complex Systems

Bar-Yam states that the "importance of complex systems ideas in technology begins through the recognition that novel technologies promise to enable us to create ever more complex systems." And he continues saying that "the conventional boundary between technology and the human beings that use them is not a useful approach to thinking about complex systems of human beings and technology" [17]. Complex systems are represented in almost every field of science. Hence this discipline is linked to all other disciplines in Fig. 5.1 and therefore is centered in the diagram. The field of complex systems itself can be subdivided into three sections: complex adaptive systems, biological systems and complex products and systems (CoPS), as shown in Fig. 5.2. These sections all mutually overlap.

Goldberg and Holland mention the "robust complexity that evolution has achieved in its three billion years of operation". And further: "The 'genetic programs' of even the simplest living organisms are more complex than the most intricate human designs" [18]. And in another publication the authors state that "Adaptive processes, with rare exceptions, are far more complex than the most complex processes studied in the physical sciences. And there

Fig. 5.2 Categories of complex
systems

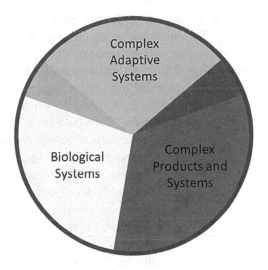

is as little hope of understanding them without the help of theory as there would be of understanding physics without the attendant theoretical framework" [19].

"A Complex Adaptive System (CAS) has no single governing equation or rule that controls the system. Instead, it has many distributed, interacting parts, with little or nothing in the way of a central control. Each of the parts is governed by its own rules. Each of these rules may participate in influencing an outcome, and each may influence the actions of other parts" [20]. All complex adaptive systems share three characteristics: evolution, aggregate behavior and anticipation [20]. Ecosystems, and the global biosphere, are prototypical examples of CASes [21].

Ecosystems and the biosphere are also biological systems. Regarding such systems, Holling mentions "that the complexity of living systems of people and nature emerges not from a random association of a large number of interacting factors rather from a smaller number of controlling processes. These systems are self-organized, and a small set of critical processes create and maintain this self-organization" [22]. He further explains that "'Self-organization' is a term that characterizes the development of complex adaptive systems, in which multiple outcomes typically are possible". Holling introduces a "theoretical framework and process for understanding complex systems" that shall be "as simple as possible but no simpler", "be dynamic and prescriptive" and "embrace uncertainty and unpredictability" [22]. Albin focuses on humans and their social interactions as biological systems and states that "The program of reducing complex physical interactions to computable models, which has had such striking success in the physical sciences, has much narrower conceptual limits in the context of social interactions, precisely because the constituent subsystems of societies, human beings, are themselves emergent, complex adaptive systems" [23]. And in this context Folke summarizes that "Not only adaptations to current conditions and in the short term, but how to achieve transformations toward more sustainable development pathways is one of the great challenges for humanity in the decades to come" [24]. In their publication titled "Complexity of Coupled Human and Natural Systems" Liu et al. discovered based on several case studies that "Some ecosystems can only be sustained through human management practices, whereas many conservation efforts preclude such human interference" [25].

Advances in technology and science resulted in designing more and more complex systems. Around the turn of the millennium scientists started to talk about Complex Products [and] Systems (CoPS) which are "high cost, engineering-intensive products, systems, networks and constructs" [26] (according to [27]). "They are business to business capital goods used to produce consumer goods and services" [26]. Considering the specific challenges of CoPS, Gann and Salter mention that "Key issues for makers of complex products and systems in the built environment are not solely the management of projects or the management of business processes per se, but rather the integration of project and business processes within the firm" [28]. The same authors highlight the importance of knowledge management on all enterprise levels for successfully manage CoPS challenges. And Hobday explains that "Technical progress, combined with new industrial demands have greatly enhanced the functional scope, complexity, pervasiveness, and performance of CoPS e.g., business information networks, tailored software packages, and internet super-servers. The nature of CoPS can lead to extreme task complexity, which, in turn, demands particular forms of management and industrial organisation" [29]. And as a further challenge the author mentions: "Complex and changing clients needs are not unusual in CoPS; therefore, a pre-emptive, pro-active approach is essential to minimising risks". In general, complex systems cannot be completely modeled and described, so measures for their management must have other targets. Booker et al. formulate that "A complex environment will contain concepts that cannot be specified easily or precisely even with a powerful logic" [19]. These authors postulate that managing complexity should reduce uncertainty. And Sterman proclaims that we have to learn in and about complex systems in order to be able to manage them. And "Overcoming the barriers to learning requires a synthesis of many methods and disciplines, from mathematics and computer science to psychology and organizational theory. Theoretical studies must be integrated with field work. Interventions in real organizations must be subjected to rigorous follow-up research" [30].

5.2.4   Software Engineering

According to ISO/IEC/IEEE, software engineering is "the systematic application of scientific and technological knowledge, methods, and experience to the design, implementation, testing, and documentation of software" [31]. In this engineering discipline, an important question is the measurement of complexity as a metric that can e.g. trigger measures of system change and optimization.

In 1976, McCabe "describes a graph-theoretic complexity measure and illustrates how it can be used to manage and control program complexity" [32]. His mathematical approach is built to "measure and control the number of paths through a program". McCabe used his approach for answering the question: "How to modularize a software system so the resulting modules are both testable and maintainable?" Thus, the objective is not only to obtain knowledge about the program complexity, but to use this knowledge for program optimization, decision-making and management.

The software engineering question of how to modularize software systems is similar to architectural questions in product development and systems engineering. Therefore, similar methods can be applied using discipline-specific objects and optimization parameters as model components. For example, interface management can be used for identifying suitable software modules when trying to streamline information flows and minimize change impact. McCabe and Butler provide three measures, "module design complexity, design complexity, and integration complexity" and explain that those are important complexity management tools [33].

Complexity of software applications increased dramatically since the early days of computer science. Thus, software programs evolved towards complex projects, which explains the overlap between modern software engineering and systems engineering (see Sect. 5.3). In several aspects, e.g. model-based engineering and testing, software engineering drives progress in systems engineering [34].

5.2.5 Management Science

Management science describes the way management faces organizations and their complexity. The organization represents the system with groups, units or departments as its subsystems. Those subsystems (and the included people) are interconnected by information flows and build a highly dynamic and complex system.

Schein shows that "organizational culture is a complex phenomenon" and tries to provide a better understanding of behavior and effects observed in organizations. However, he summarizes that there is still a lack of understanding concerning many aspects of this organizational complexity [35]. Stacey links the management of organizations with complexity as follows: "First, why should organizational theorists pay attention to the science of complexity? The answer is that organizations are nonlinear, network feedback systems and it therefore follows logically that the fundamental properties of such systems should apply to organizations." The author further states that "Organizations are clearly feedback systems because every time two humans interact with each other the actions of one person have consequences for the other, leading that other to react in ways that have consequences for the first, requiring in turn a response from the first and so on through time." And the "long-term outcomes [from interventions conducted by members of the organization] emerge from a process which is basically self-organizing" [36]. Stacey proclaims to adopt findings made from managing complex systems in other disciplines to better manage organizations. So, he explains that when organizations operate in a stable equilibrium, where they can change in predictable ways to different equilibriums then the shift to another stable state is difficult, because the organization and its members try to stay in the original equilibrium. This behavior is not fruitful for innovation or creativity, because the outcome of these changes is unforeseeable. Stacey concluded that "The science of complexity demonstrates that for a system to be innovative, creative, and changeable it must be driven far from equilibrium where it can make use of disorder, irregularity, and difference as essential elements in the process of change" [36].

5.2.6 Complexity Theory

The field of complexity theory aggregates the scientific approaches towards a general understanding and modeling of complexity. These approaches are inspired by observations from nature and represent the starting point for managing complexity in various fields of application. The historical aspects of developing complexity theories are described in detail in Chap. 4. Here, recent aspects of general complexity understanding are indicated.

In 1995, Kauffman stated that "The past three centuries of science have been predominantly reductionist, attempting to break complex systems into simple parts, and those parts, in turn, into simpler parts". And furthermore: "How do we use the information gleaned about the parts to build up a theory of the whole? The deep difficulty here lies in the fact that the complex whole may exhibit properties that are not readily explained by understanding the parts. The complex whole, in a completely nonmystical sense, can often exhibit collective properties, 'emergent' features that are lawful in their own right" [37].

It is interesting to see how the reductionistic approach towards complexity was revived several times after its initial creation (see also Sect. 4.1.2). And even recent definitions and perspectives in the engineering context show a partly reductionistic focus on system decomposition into distinct components (see Sect. 3.2). Obviously, the success of this approach towards complexity is based on its successful application. However, it only works for a subset of complexity challenges, and Kauffman mentions "The reductionist program has been spectacularly successful, and will continue to be so. But it has often left a vacuum" [37].

Kauffman mentions that from the twentieth century on, science is confronted with organized complexity and "nowhere is this confrontation so stark as in biology" [38]. Other authors connected complexity to topics like economics [39]. He also describes a concept of inductive reasoning, which is applied by humans when interacting with complex systems or in complex environments [40]. This approach is used for agents in financial markets as well.

McKelvey refers to thermodynamics and evolutionary processes in his publication titled "What is complexity science? It is really order-creation science". He criticized that "attention to the basic causal process underlying emergence has largely been ignored in most managerial and organizational applications of complexity science". He claims that "The classic concept of external (Bénard) energy differentials (as control parameters) that cause emergence 'at the edge of chaos' is at the heart of complexity science, but is frequently missing in much of the complexity science literature and particularly in organizational applications" [41].

Kauffmann also uses thermodynamic analogies for describing aspects of complexity theory: "when water freezes, one does not know where every water molecule is, but a lot can be said about your typical lump of ice. It has a characteristic temperature, color, and hardness—'robust' or 'generic' features that do not depend on the details of its construction. And so it might be with complex systems such as organisms and economies. Not knowing the details we nevertheless can build theories that seek to explain the generic properties'" [37].

5.2.7 Complex Networks

Complex networks represent networks with non-trivial topological features. And typically, the investigation of complex networks has fallen into the field of graph theory [42]. Complex networks contain vertices and edges. They occur in all fields of complexity, irrespective of societal, biological or engineering systems.

The World Wide Web is a famous and supposedly well-known example. Another exemplary, complex network has been described by Barabási and Albert and is "formed by the citation patterns of the scientific publications, the vertices standing for papers published in refereed journals, the edges representing links to the articles cited in a paper" [43]. A similar project is the so called Erdős number, indicating a scientist's distance to Paul Erdős, measured by (co-)authorship of mathematical publications [44].

There is a great number of networks surrounding us. Most of them are complex even if their structure can be described in one sentence as the example above has shown. Jeong et al. describe in their publication titled "Error and attack tolerance of complex networks" that "Many complex systems display a surprising degree of tolerance against errors. For example, relatively simple organisms grow, persist and reproduce despite drastic pharmaceutical or environmental interventions, an error tolerance attributed to the robustness of the underlying metabolic network. Complex communication networks display a surprising degree of robustness: while key components regularly malfunction, local failures rarely lead to the loss of the global information-carrying ability of the network" [45]. Although there are innumerous attacks against computers, or in other words against vertices of the World Wide Web, the network itself is unaffected and still fully operative.

5.3 Discipline Overlaps

Besides the seven fields of engineering complexity, Fig. 5.1 shows eleven overlaps between these fields. The field of complex systems is centrally located and is involved in most of these overlaps. That seems to be understandable, as complex systems play an important role in almost any engineering domain. Figure 5.1 further indicates overlaps between the fields of complexity theory and complex networks, product development and system engineering, as well as systems engineering and software engineering. For these overlaps, the following sections shall briefly mention selected authors and some of their significant statements that contribute to research in those areas.

I: Complex Systems and Systems Engineering

Several research works can be placed in the intersection of the two fields complex systems and systems engineering. Bar-Yam analyzes early success stories of systems engineering, the Manhattan Project and space projects in the US, and compares them with more recent projects, e.g. the modernization of the US air traffic control. Assessing the success of the old pioneering projects of systems engineering he states that "Many projects end up as failed and abandoned. This is true despite the tremendous investments that are made". And when looking into more recent projects he concludes that "A fundamental reason for the difficulties with modern large engineering projects is their inherent complexity. Complexity is generally a characteristic of large engineering projects today. Complexity implies that different parts of the system are interdependent so that changes in one part may have effects on other parts of the system. Complexity may cause unanticipated effects that lead to failures of the system". Bar-Yam then mentions feedback loop models as a possible approach for managing those undesired effects [3].

Furthermore, Bar-Yam states that "While the complexity of engineering projects has been increasing, it is important to recognize that complexity is not new". And he refers to the existing approaches towards complexity management modularity and abstraction as being "useful, but at some degree of interdependence [. . .] become ineffective". He claims that a "concept of incremental design is one step towards a more complex systems oriented approach" for modern, complex projects. Therefore, Bar-Yam proposes "that complex engineering projects should be managed as evolutionary processes that undergo continuous rapid improvement through iterative incremental changes performed in parallel and thus is linked to diverse small subsystems of various sizes and relationships" [3].

Abbott states that "it has become increasingly clear that systems engineering is the engineering of complex systems" and "although complex systems ideas, tools, and techniques have been applied to systems engineering problems for some time, there has been little effort to date to bring the two fields [. . .] together in a more formal and explicit way". Abbott describes the approach towards installing a working group with workshops held at conferences from both research fields with the objective "to introduce these communities to each other and to embark on what we expect to be an extended dialog". Abbott presents an initial catalog of 28 areas of interest for bringing the two fields closer together and planned to initiate discussions within the structures of the associations IEEE and INCOSE [15].

Norman and Kuras "introduce a new set of processes which complement—and do not replace—the processes that constitute traditional systems engineering" and call the result complex systems engineering. The authors formulate the motivation for this evolution of systems engineering as follows: "Among the characteristics one would require to have a successful, or at least a low risk outcome, there are a few which are absolutely required to ensure success using traditional Systems Engineering. These serve as boundary conditions for applying T[raditional]S[ystems]E[ngineering]". These boundary conditions, the desired outcome known a priori, a central resource allocation and change management, as well as "fungible" resources are violated when dealing with a complex system like an enterprise [46].

Table 5.1 Comparing traditional systems engineering and complex systems engineering, according to [46]

Traditional systems engineering	Complex systems engineering
Products are reproducible	No two enterprises are alike
Products are realized to meet pre-conceived specifications	Enterprises continually evolve so as to increase their own complexity
Products have well-defined boundaries	Enterprises have ambiguous boundaries
Unwanted possibilities are removed during the realizations of products	New possibilities are constantly assessed for utility and feasibility in the evolution of an enterprise
External agents integrate products	Enterprises are self-integrating and re-integrating
Development always ends for each instance of product realization	Enterprise development never ends— enterprises evolve
Product development ends when unwanted possibilities are removed and sources of internal friction (competition for resources, differing interpretations of the same inputs, etc.) are removed	Enterprises depend on both internal cooperation and internal competition to stimulate their evolution

Norman and Kuras state that traditional systems engineering methods do not scale in order to be applicable to complex systems and therefore propose a simultaneous application: "Traditional and complex system engineering can (and should) be applied concurrently in the realization and evolution of a complex system. Traditional system engineering is appropriate for managing the decision making processes of individual autonomous agents in a complex system. Complex system engineering must be added when multiple autonomous agents must be a part of any solution and/or when multiscale analysis becomes essential to a sufficiently complete characterization of an evolving problem and its solution" [46].

Table 5.1 shows a high-level comparison of traditional systems engineering and complex systems engineering as seen by Norman and Kuras. In their publication, the authors also provide a detailed description of a complex systems engineering application in a large-scale military project.

In her publication "Bridging Systems Engineering and Complex Systems Sciences", Sheard is investigating complex software projects and states that "Systems engineering, meaning our ability to engineer increasingly more complex systems, is in crisis at the start of this third millennium". She sees the reason therefore in increasing complexity and proposes a closer connection between the fields of systems engineering and complex systems. "Complex systems are discovering principles that directly apply in many ways to the problems of systems engineering, yet for the most part the bridge between these systems sciences and systems engineering is inadequate". Sheard describes her view on how different systems engineering stakeholders are familiar with complex systems and how knowledge and methods from the field of complex systems could help deal with principles of systems engineering [16].

II: Complex Systems and Product Development

Amari et al. consider complex systems from a product development perspective. The authors propose two objective complexity measures to be applied to product design. They mention the importance of having clear knowledge about design complexity for three main reasons: "First, it helps design engineers to develop a better understanding of various aspects of complexity thereby evolving toward simpler design solutions. Second, it enables design automation tools to systematically evaluate different design alternatives based on their inherent complexities. [...] Finally, it provides engineering design researchers with a theoretical framework for rigorous and unambiguous characterization of complexity in design" [47].

Craig Read also highlights that "Complexity is a significant factor in the development of new products and systems". He identifies aspects from a product development view that are linked to complexity from a system perspective. These aspects are "interoperability; upgradability; adaptability; evolving requirements; system size; automation requirements; performance requirements; support requirements; sustainability; reliability; the need for increased product lifespan; and finally, the length of time systems take to develop". Read aims at developing a better "understanding of systems complexity" by describing "complexity within engineered systems" [1].

Sharman and Yassine address complex product architectures and present "three methods for describing product architectures". Their objective is to "describe and grasp the structure of the product" and to "facilitate[e] product modularization". The authors indicate that complex product architectures require methods of abstraction in order to become manageable. And they state that the three methods they introduce "can be used to qualitatively or quantitatively characterize any given architecture spanning the modular-integrated continuum" [48].

III: Complex Systems and Complexity Theory

The overlap of the two fields complex systems and complexity theory seems to be obligatory. Findings from observing complex systems build the basis for research in the field of complexity theory. And the scientific approaches towards understanding and modeling complexity then can be applied to model, understand and interact with complex systems. Thus, principles of self-organization were observed in complex biological systems, which became an important research topic in complexity theory and get applied for modeling complex systems.

When modeling complex adaptive systems, basic aspects from the fields of complexity theory and complex systems get applied [37]. According to Holland, "Cas are systems that have a large numbers of components, often called agents, that interact and adapt or learn" [49]. He further declares that "many difficult contemporary problems center on complex adaptive systems" and introduces the following exemplary list of problems, which can be modeled as complex adaptive systems: encouraging innovation in dynamic economies, providing for sustainable human growth, predicting changes in global trade, understanding markets, preserving ecosystems, controlling the Internet (e.g. controlling viruses and spam) and strengthening the immune system [49].

Holland introduces that all complex adaptive systems "share four major features": Parallelism, conditional action, modularity and adaption and evolution [49]. Dealing with these features can be supported by approaches from complexity theory and methods and procedures applied to manage complex systems.

Rosser applies an approach of dynamic complexity for explaining economic phenomena [50]. Especially the "phenomenon of emergence, the appearance of new forms or structures at higher levels of a system from processes occurring at lower levels" is located in the field of complexity theory [50]. The computational means then can be anchored to concepts in the field of complex systems.

IV: Complex Systems and Complex Networks
Many approaches of modeling and visualizing complex systems apply possibilities emerging from the field of complex networks. With findings from the field of complex networks, identified or observed systems can be acquired and depicted. And new observations can motivate new approaches of network modeling. The mathematical fundaments of graph theory provide powerful means for describing, analyzing and interacting with networks representing complex systems.

For example, Albert and Barabasi propose a method that "can serve as a roadmap for understanding the dynamics of large interacting systems in general", applying Boolean networks for modeling complex networks [51]. In 2001, the same authors provide a comprehensive introduction to different types of systems forming complex networks in their publication "Statistical Mechanics of Complex Networks" [42]. For visualizing the interconnectivity between complex system elements, graph representations are the dominant approach. Due to the availability of high computational power and associated dynamic representations, a large variety of graphs are in use for representing even large-scale complex systems with several attributes in networked format. Lima provides a comprehensive overview of network types with many examples [52].

V: Complex Systems and Management Science
Because solving economic challenges was among the early applications of system thinking and cybernetics, it is not surprising to see strong overlaps between the fields of complex systems and management sciences. Many formal management approaches apply techniques, methods and modeling forms taken and adapted from work in complex systems.

In management science, decision-making and learning how to come to optimal decisions in dynamic environments is of major relevance. Several authors describe the necessity of training possibilities for managers comparable to a flight simulator for pilots, e.g. Senge and Sterman. These authors further mention the importance of organizational learning for managers in dynamic enterprises and propose a simulation-based learning laboratory [53]. With such a laboratory they want to allow more rapid learning and increased flexibility in a world of growing complexity and change. They also state that "For systems theorists, the source of poor performance and organizational failure is often to be found in the limited cognitive skills and capabilities of individuals compared to the complexity of the systems they are called upon to manage". And that "Dynamic decision making is particularly difficult, especially when decisions have indirect, delayed, nonlinear, and multiple feedback effects" [53]. For these statements, Senge and Sterman refer to earlier publications dealing with topics at the intersection of research fields, i.e. by Jay Forrester and Dietrich Dörner [54, 55].

Management science deals with human beings and their limited capability of understanding complex systems. Diehl and Sterman describe an experiment where test subjects had to make decisions in a dynamic, complex environment. The test subjects' "performance deteriorated dramatically with increasing time delays and feedback effects" despite perfect knowledge of the system's structure and its parameters [56].

Uhl-Bien et al. focus on leadership and state that "complexity science suggests a different paradigm for leadership". They "develop an overarching framework for the study of Complexity Leadership Theory, a leadership paradigm that focuses on enabling the learning, creative, and adaptive capacity of complex adaptive systems (CAS) within a context of knowledge-producing organizations". The framework comprises three leadership functions—adaptive, administrative and enabling leadership—and are dynamically "intertwined". The authors conclude that "leadership is too complex to be described as only the act of an individual or individuals; rather, it is a complex interplay of many interacting forces" [57].

In modern industries the management of supply networks became an increasingly complex challenge. Choi et al. describe how "managers have struggled with the dynamic and complex nature of supply networks (SNs) and the inevitable lack of prediction and control". They explain that "in the current literature, a deterministic or deliberate approach to managing the SN has been emphasized" and why such an approach "may be effective only to a certain extent" and "may eventually stagnate". As an alternative to conventional supply chain management, Choi et al. highlight "the need to recognize supply networks as a complex adaptive system", to apply concepts and principles of complex adaptive systems to supply networks management and to discuss the consequences [58].

Project scheduling represents a management task that can become a complex challenge when a large number of interlinked project activities have to be managed. Many approaches for measuring the complexity of such activity networks have been developed, mainly depending on the number of network nodes and arcs/edges connecting them. Some advanced approaches make use of means of graph theory and also take into account the embedding of activities into the network. For example, such approaches determine the number of preceding and succeeding activities or the adjacency between nodes, and use this information to deduce the behavior of the networks. A summary of network complexity measures can be seen in [59].

In the context of project planning, Lévárdy and Browning mention that conventional project planning with its a priori specification and scheduling of project activities can be inadequate when dealing with highly complex projects. Therefore, they propose an adaptive product development process (APDP) modeling framework "that views the PD [product development] process as a complex adaptive system". "Rather than presuming that a particular set of activities and interactions is necessary and sufficient to achieve a project's goal, the model accounts more broadly for a superset of potentially relevant activity modes and interactions. From this "primordial soup," we explore what types of processes emerge and their comparative fitness (or value, in terms of risk reduction) in achieving the goals" [60].

VI: Complex Systems and Software Engineering

Since the beginning of software development its complexity has increased tremendously. In many product systems the majority of functionalities is realized by software. And not only software modules, but also software-enabled products interact in increasingly complex systems. Besides questions of modularity and integral design, larger software projects have to deal with additional complexity in additional system views, e.g. process networks and organizational structures.

The increasing size of software systems made it necessary to adopt findings from the field of complex systems in software engineering. For example, Jennings mentions that "Agents are being advocated as a next generation model for engineering complex, distributed [software] systems" [61]. In the publication titled "An agent-based approach for building complex software systems", Jennings addresses explicitly the need for complexity management when developing software systems and states that "Industrial-strength software is complex: it has a large number of parts that have many interactions [...]. Moreover this complexity is not accidental [...], it is an innate property of large systems. Given this situation, the role of software engineering is to provide structures and techniques that make it easier to handle complexity" [61]. In addition to this systems perspective on software engineering Jennings describes decomposition, abstraction and modularization (Jennings calls this "organization") as "fundamental tools [...] for helping to manage complexity". This is congruent with a structural description of complex systems.

VII: Systems Engineering and Software Engineering

Also the fields of systems engineering and software engineering show many commonalities. In fact, nowadays both fields are widely merged, so that software engineering applications make a significant part of the contributions to systems engineering conferences and journals (see e.g. contributions made to the INCOSE Symposium or the journal "Systems Engineering").

However, Boehm explains that software engineering and systems engineering started with different premises and the trend towards "increasing integration of software engineering and systems engineering" is a more recent development [62]. He describes that "systems engineering began as a discipline for determining how best to configure various hardware components into physical systems [...]. Once the systems were configured and their component functional and interface requirements were precisely specified, sequential external or internal contracts could be defined for producing the components. When software components began to appear in such systems, the natural thing to do was to treat them sequentially and independently as Computer Software Configuration Items" [62]. Boehm describes that in the beginning of software engineering a "reductionist" development of software components was focused. With projects becoming more software-intensive, "software people were recognizing that their sequential, reductionist processes were not conducive to producing user-satisfactory software, and were developing alternative software engineering processes (evolutionary, spiral, agile) involving more and more systems engineering activities. Concurrently, systems engineering people were coming to similar conclusions about their sequential, reductionist processes, and developing alternative "soft systems engineering" processes" [62].

VIII: Systems Engineering and Product Development
The development of a technical product comprises part of many systems engineering projects. Thus, an overlap between both fields in terms of complexity management suggests itself. Approaches and methods of product development can be applied in several parts of the systems engineering process, and they can be adopted and transferred to others. For example, modularization can not only be used for product structure optimization, but also for associating tasks to organizational units.

On the other hand, strategies of complexity management originating from application in the systems engineering process can be helpful in the specific context of product development. Information and risk management can be named by examples. From a systems engineering perspective on product development, Browning et al. state that "Progress is made and value is added by creating useful information that reduces uncertainty and/or ambiguity. But it is challenging to produce information at the right time, when it will be most useful. Developing complex and/or novel systems multiplies these challenges" [63]. The authors state "that making progress and adding customer value in P[roduct]D [evelopment] equate[s] with producing useful information that reduces performance risk". Therefore they propose an approach that "integrates several concepts and methods, including technical performance measures (TPMs), risk reduction profiles, customer preferences, and uncertainty" [63].

Most definitions of systems engineering mention risk management as a key element for successful system creation. Browning and Eppinger state that "firms that design and develop complex products seek to increase the efficiency and predictability of their development processes" for gaining "competitive leverage". They model the product development process including several characteristics and create a possibility to compare "Alternative process architectures [...] revealing opportunities to trade cost and schedule risk" [64].

Browning highlights that "A process, as a kind of system, derives its added value from the relationships among its elements (e.g. activities)" [65]. He emphasizes that engineering processes are "especially complex because of the large number of interdependencies among the activities" and that "The systems engineering 'V' model applies to processes as well as to products". Within this systems engineering context, Browning presents the design structure matrix as "a powerful technique for representing and analyzing complex processes" [65]. Backed by means of operations research, Ahmadi et al. present a similar approach for structuring product development processes [66].

Browning et al. investigate "process modeling in a systems engineering context" and mention that "while product systems must be created, the process systems for developing complex products must, to a greater extent, be discovered and induced". The authors present important concepts and a framework for modeling product development processes [67].

IX: Complexity Theory and Complex Networks

While the field of complexity theory comprises the scientific approaches towards understanding of complexity, the field of complex networks investigates possibilities of modeling complex systems. This modeling is often assigned to graph theory. The depiction of conventional, static networks consisting of nodes and edges is well-established and can be enriched with the modeling of additional parameters. This serves many applications of a reductionistic complexity model, e.g. decomposition, modularization or integration approaches.

Complexity characteristics like dynamics and self-organization require more enhanced modeling possibilities than a simple node-and-edge diagram can provide. Anderson describes how close progress in computing has been linked to modeling and studying complex systems. He specifically mentions cellular automata, neural networks and genetic algorithms as powerful approaches [68].

X: Complex Systems and Complexity Theory and Complex Networks

The overlap of the three fields is best explained with a specific example. Merali and McKelvey call their research complexity science, and describe it "as a source of concepts for enabling the trans-disciplinary exploration of complex organization in the network economy and network society, and for explaining the dynamics of networked systems at different levels of description ranging from the micro- to the macro-level" [69].

One can argue that such new concepts belong to the field of complexity theory, where scientific approaches towards the understanding of complexity are investigated. Complex organizations can be observed in the field of complex systems. And the consideration of such systems by network approaches belongs to the field of complex networks.

XI: Complex Systems and Systems Engineering and Product Development

While the mutual overlaps between the fields of systems engineering, product development and complex systems have been described above, a common merging of all three fields can also be argued. For example, concurrent engineering approaches can be seen applied in the field of systems engineering as well as in specific applications to (technical) product development. Yassine and Braha propose a complex system modeling approach for such concurrent engineering [70]. Specifically, the authors propose the application of design structure matrices, which are useful for depicting and analyzing static networks.

5.4   Discussion

The classification of complexity management into seven research fields in this chapter is one approach of structuring this tremendously large and steadily growing field. And the indicated overlaps do not necessarily represent the only merging points where research has been done. Especially, the application of different vocabulary, terms and definitions can make it difficult to see similarities, correlations, but also transferability between approaches.

Several authors stated that managing a complex system is a complex challenge in itself—with dynamic system behavior as one of its characteristics. The same accounts for the classification of complexity management research in engineering. So, the descriptions in this chapter can only represent a starting point and guideline for further perspectives and specific investigations. The graphic depiction of the Venn diagram in Fig. 5.1 serves the additional purpose of easy accessibility to the classification. This facilitates discussions, criticism and improvement of the classification.

The contributions mentioned in the categories and their overlaps are simply examples that were chosen, and many more works could easily be added in all areas. Nevertheless, the indicated publications represent useful starting points for acquiring knowledge about specific fields and relevant topics.

As mentioned above, categorizing the fields of engineering complexity is a complex challenge comprising dynamic behavior. And as they did in the past, the challenges, technical possibilities and associated approaches will change in the future. Therefore it is an interesting task to think through to the future importance and development of the categories and overlaps. For example, the trend in the field of systems engineering towards the consideration of increasingly comprehensive and complex systems of systems may suggest that this category will adopt many approaches from other categories in the future.

Besides questioning the categories and overlaps depicted in Fig. 5.1, the consideration of possible overlaps in general is interesting—which have not been indicated in this chapter. Basically, an overlap of categories means the transfer of expertise from one field of research or application to another. And besides a mere adoption, the transfer of expertise does often also represent the origin of new approaches.

The main objective of the descriptions in this chapter is to provide closer insight into the diversified field of complexity management research in engineering. And this helps in reducing the lack of knowledge about complexity, its challenges and possibilities to manage—as it has been highlighted by many authors in the past.

References

1. Read, Craig. 2008. Complexity Characteristics and Measurement within Engineering Systems.
2. Sheard, Sarah A, and Ali Mostashari. 2010. A Complexity Typology for Systems Engineering.
3. Bar-Yam, Y. 2003. When Systems Engineering Fails-toward Complex Systems Engineering. SMC'03 Conference Proceedings. 2003 I.E. International Conference on Systems, Man and Cybernetics. Conference Theme—System Security and Assurance (Cat. No.03CH37483) 2. IEEE: 2021–28. doi:10.1109/ICSMC.2003.1244709.
4. Eppinger, Steven D. 2002. Patterns of Product Development Interactions.
5. Conway, M. 1968. How Do Committees Invent? *Datamation* 14: 28–31.
6. Ulrich, Karl. 1995. The Role of Product Architecture in the Manufacturing Firm. *Research Policy* 24(3): 419–440. doi:10.1016/0048-7333(94)00775-3.
7. Eppinger, Steven D., and Tyson R. Browning. 2012. *Design Structure Matrix Methods and Applications*. Cambridge, MA: MIT Press.
8. Puls, C., L. Bongulielmi, P. Henseler, and M. Meier. 2002. Management of Different Types of Configuration Knowledge with the K- & V-Matrix and Wiki. In Proceedings of the 7th International Design Conference 2002 (DESIGN02), ed. D. Marjanovic. Cavtat-Dubrovnik: Design Society.
9. Yassine, Ali, Whitney Daniel, S. Daleiden, and J. Lavine. 2003. Connectivity Maps: Modeling and Analysing Relationships in Product Development Processes. *Journal of Engineering Design* 14(3): 377–394.
10. Lindemann, Udo, Maik Maurer, and Thomas Braun. 2009. *Structural Complexity Management—An Approach for the Field of Product Design*. Berlin. Springer http://medcontent.metapress.com/index/A65RM03P4874243N.pdf.
11. SEBoK—Guide to the Systems Engineering Body of Knowledge. 2015. Systems Engineering: Historic and Future Challenges. http://sebokwiki.org/wiki/Systems_Engineering:_Historic_and_Future_Challenges.
12. Blanchard, B.S. 2008. *System Engineering Management*. 4th ed. Hoboken: Wiley.
13. Kasser, Joe. 1996. Systems Engineering: Myth or Reality? In INCOSE International Symposium, 877–81. Wiley Online Library.
14. Sheard, SA. 2000. Three Types of Systems Engineering Implementation. Proceedings of the INCOSE, no. July. ftp://76.171.69.51/Resources/Systems Engineering/ENM 607A/Resources III/Three Types of Systems Engineering.pdf.
15. Abbott, Russ. 2006. Complex Systems + Systems Engineering = Complex Systems Engineering.
16. Sheard, Sarah. 2006. Bridging Systems Engineering and Complex Systems Sciences, 1–7. Third Millennium Systems LLC.
17. Bar-Yam, Yaneer. 2003. *Unifying Principles in Complex Systems*.
18. Goldberg, D.E., and John H. Holland. 1988. Genetic Algorithms and Machine Learning, 95–99.
19. Booker, L.B., D.E. Goldberg, and J.H. Holland. 1989. Classifier Systems and Genetic Algorithms. *Artificial Intelligence* 40(1–3): 235–282. doi:10.1016/0004-3702(89)90050-7.
20. Holland, John H. 1992. Complex Adaptive Systems.
21. Levin, Simon A. 1998. Ecosystems and the Biosphere as Complex Adaptive Systems. *Ecosystems* 1: 431–436.
22. Holling, C.S. 2001. Understanding the Complexity of Economic, Ecological, and Social Systems. *Ecosystems* 4(5): 390–405. doi:10.1007/s10021-001-0101-5.
23. Albin, Peter S. 1998. *Barriers and Bounds to Rationality: Essays on Economic Complexity and Dynamics in Interactive Systems*. Princeton, NJ: Princeton University Press.
24. Folke, Carl. 2006. Resilience: The Emergence of a Perspective for Social–Ecological Systems Analyses. *Global Environmental Change* 16(3): 253–267. doi:10.1016/j.gloenvcha.2006.04.002.

25. Liu, Jianguo, Thomas Dietz, Stephen R. Carpenter, Marina Alberti, Carl Folke, Emilio Moran, Alice N. Pell, et al. 2007. Complexity of Coupled Human and Natural Systems. *Science* 317 (5844): 1513–1516. doi:10.1126/science.1144004.
26. Ren, Ying-Tao, and Khim-Teck Yeo. 2006. Research Challenges on Complex Product Systems (CoPS) Innovation. *Journal of the Chinese Institute of Industrial Engineers* 23(6): 519–529. doi:10.1080/10170660609509348.
27. Hobday, Mike. 1998. Product Complexity, Innovation and Industrial Organisation. *Research Policy* 26: 689–710.
28. Gann, David M., and Ammon J. Salter. 2000. Innovation in Project-Based, Service-Enhanced Firms: The Construction of Complex Products and Systems. *Research Policy* 29(7–8): 955–972. doi:10.1016/S0048-7333(00)00114-1.
29. Hobday, Mike. 2000. The Project-Based Organisation: An Ideal Form for Managing Complex Products and Systems? *Research Policy* 29(7–8): 871–893. doi:10.1016/S0048-7333(00)00110-4.
30. Sterman, John D. 1994. Learning in and about Complex Systems. *System Dynamics Review* 10 (February): 291–330.
31. Systems and Software Engineering—Vocabulary. 2010.
32. McCabe, Thomas J. 1976. A Complexity Measure. *IEEE Transactions on Software Engineering* SE-2(4): 308–320.
33. McCabe, Thomas J., and Charles W. Butler. 1989. Design Complexity Measurement and Testing. *Communications of the ACM* 32(12): 1415–1425. doi:10.1145/76380.76382.
34. Ogren, Ingmar. 2000. On Principles for Model-Based Systems Engineering. *System Engineering* 3: 38–49. doi:10.1002/(SICI)1520-6858(2000)3:1<38::AID-SYS3>3.0.CO;2-B.
35. Schein, Edgar H. 1990. Organizational Culture. *American Psychologist* 45(2): 109–119. doi:10.1037/0003-066X/90/S00.75.
36. Stacey, Ralph D. 1995. The Science of Complexity: An Alternative Perspective for Strategic Change Processes. *Strategic Management Journal* 16(6): 477–495.
37. Kauffman, Stuart. 1995. *At Home in the Universe*. New York: Oxford University Press.
38. ———. 1993. *The Origins of Order*. New York: Oxford University Press.
39. Arthur, W. Brian. 1995. Complexity in Economic and Financial Markets. *Complexity* 1(1).
40. ———. 1994. Inductive Reasoning and Bounded Rationality.
41. McKelvey, Bill. 2001. What is Complexity Science? *Emergence* 3(1): 137–157.
42. Albert, Réka, and Albert-László Barabási. 2001. Statistical Mechanics of Complex Networks.
43. Barabási, Albert-László, and Réka Albert. 1999. Emergence of Scaling in Random Networks, 1–11 Department of Physics, University of Notre-Dame, Notre-Dame, IN 46556 Systems.
44. The Erdös Number Project. 2015. http://wwwp.oakland.edu/enp/. Accessed 29 Dec.
45. Jeong, Hawoong, Réka Albert, and Albert-László Barabási. 2000. Error and Attack Tolerance of Complex Networks, 1–14.
46. Norman, Douglas O., and Michael L. Kuras. 2006. Chapter α Engineering Complex Systems.
47. Ameri, Farhad, Joshua D. Summers, Gregory M. Mocko, and Matthew Porter. 2008. Engineering Design Complexity: An Investigation of Methods and Measures. *Research in Engineering Design* 19(2–3): 161–179. doi:10.1007/s00163-008-0053-2.
48. Sharman, David M., and Ali A. Yassine. 2004. Characterizing Complex Product Architectures. *Systems Engineering* 7: 35–60. doi:10.1002/sys.10056.
49. Holland, John H. 2006. Studying Complex Adaptive Systems, no. November 2005, 1–8.
50. Rosser, J. Barkley. 2006. Dynamic and Computational Complexity in Economics, no. June, 1–21.
51. Albert, R., and Al Barabasi. 2000. Dynamics of Complex Systems: Scaling Laws for the Period of Boolean Networks. *Physical Review Letters* 84(24): 5660–5663 .http://www.ncbi.nlm.nih.gov/pubmed/10991019

52. Lima, Manuel. 2011. *Visual Complexity—Mapping Patterns of Information*. New York: Princeton Architectural Press.
53. Senge, Peter M., and John D. Sterman. 1992. Systems Thinking and Organizational Learning: Acting Locally and Thinking Globally in the Organization of the Future. *European Journal of Operational Research* 59(1): 137–150. doi:10.1016/0377-2217(92)90011-W.
54. Dörner, Dietrich. 1989. Managing a Simple Ecological System. Bamberg.
55. Forrester, Jay W. 1963. *Industrial Dynamics*. Cambridge, MA: MIT Press.
56. Diehl, E., and John D. Sterman. 1995. Effects of Feedback Complexity on Dynamic Decision Making. *Organizational Behavior and Human Decision Processes* 62(2): 198–215.
57. Uhl-Bien, Mary, Russ Marion, and Bill McKelvey. 2007. Complexity Leadership Theory: Shifting Leadership from the Industrial Age to the Knowledge Era. *Leadership Quarterly* 18 (4): 298–318. doi:10.1016/j.leaqua.2007.04.002.
58. Choi, Thomas Y., Kevin J. Dooley, and Manus Rungtusanatham. 2001. Supply Networks and Complex Adaptive Systems: Control versus Emergence. *Journal of Operations Management* 19: 351–366.
59. Browning, Tyson R., and Ali A. Yassine. 2010. A Random Generator of Resource-Constrained Multi-Project Network Problems. *Journal of Scheduling* 13(2): 143–161. doi:10.1007/s10951-009-0131-y.
60. Levardy, Viktor, and Tyson R. Browning. 2009. An Adaptive Process Model to Support Product Development Project Management. *IEEE Transactions on Engineering Management* 56(4): 600–620. doi:10.1109/TEM.2009.2033144.
61. Jennings, Nicholas R. 2000. On Agent-Based Software Engineering. *Artificial Intelligence* 117 (September 1999): 277–296.
62. Boehm, Barry. 2006. Some Future Trends and Implications for Systems and Software Engineering Processes. *Systems Engineering* 9(1): 1–19. doi:10.1002/sys.20044.
63. Browning, Tyson R., John J. Deyst, Steven D. Eppinger, Daniel E. Whitney, and Senior Member. 2002. Adding Value in Product Development by Creating Information and Reducing Risk. *IEEE Transactions on Engineering Management* 49 (4):443–458.
64. Browning, T.R., and S.D. Eppinger. 2002. Modeling Impacts of Process Architecture on Cost and Schedule Risk in Product Development. *IEEE Transactions on Engineering Management* 49(4): 428–442. doi:10.1109/TEM.2002.806709.
65. Browning, Tyson R. 2002. Process Integration Using the Design Structure Matrix. *System Engineering* 5(3): 180–193.
66. Ahmadi, Reza, Thomas A. Roemer, and Robert H. Wang. 2001. Structuring Product Development Processes. *European Journal of Operational Research* 130(3): 539–558. doi:10.1016/S0377-2217(99)00412-9.
67. Browning, Tyson R., Ernst Fricke, and Herbert Negele. 2006. Key Concepts in Modeling Product Development Processes. *Systems Engineering* 9(2): 104–128. doi:10.1002/sys.20047.
68. Anderson, Philip. 1999. Complexity Theory and Organization Science. *Organization Science* 10 (3): 216–232.
69. Merali, Yasmin, and Bill McKelvey. 2006. Using Complexity Science to Effect a Paradigm Shift in Information Systems for the 21st Century. *Journal of Information Technology* 21(4): 211–215. doi:10.1057/palgrave.jit.2000082.
70. Yassine, Ali, and Dan Braha. 2003. Complex Concurrent Engineering and the Design Structure Matrix Method. *Concurrent Engineering* 11(3): 165–176. doi:10.1177/106329303034503.

A Complexity Management Framework

6

The previous chapters of this thesis provide an overview of engineering practice, historic development and the current scientific classification of complexity management approaches. But when one is confronted with a specific complexity challenge, which approach should one apply? The description of the historic development especially indicates a trend towards increasing system complexity—and with it more sophisticated models for its management. But does this mean that a classic reductionistic system model is no longer suitable for solving complex challenges? It depends, of course, on the situation.

A systematic approach towards complexity management is crucial, because complex challenges are typically characterized by a lack of clarity. But such an unclear initial situation does not allow for well-founded selection of measures to overcome complexity. In addition, complex challenges often come with high urgency for action, which can mislead people to take quick measures without assuring their suitability to the situation.

Successful complexity management requires identification of the underlying causes. Otherwise, there is a high risk of treating only the consequences of complexity, which then does not lead to a sustainable solution. It is important to clearly define the system under consideration, which means determining the system's boundaries, included elements and interdependencies between them. Such a formal definition helps to identify the type of complexity at hand, if it is about internal or external complexity and if it originates from the market, product, process or organizational area. The system definition also enables the determination of whether the considered complexity is useful or useless. Based on all these clarifications, suitable strategies and methods for handling complexity can be selected.

Figure 6.1 shows six sequential steps, which can be used as a guideline for implementing an adequate method for managing a complex challenge. This guideline describes the general logical sequence of steps, but does not represent a strict process. With each step through the guideline one obtains more information about the considered complex system. This knowledge can also require iterations of previous steps. For

© Springer-Verlag GmbH Germany 2017
M. Maurer, *Complexity Management in Engineering Design – a Primer*,
DOI 10.1007/978-3-662-53448-9_6

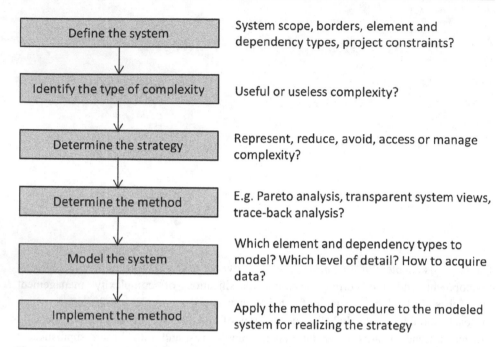

Fig. 6.1 Systematic approach on implementing complexity management for a system

example, it may become necessary to improve the system definition once the system modeling reveals ambiguities or missing clarity of element classification.

The guideline starts with the task of defining the system. This is required for identifying the origin of observed complexity. Next the type of complexity needs to be determined, e.g. if it is useful or useless in terms of the higher management objective. Before selecting a specific complexity management method the strategy has to be determined. This can be based on the type of complexity and the time scope of the management approach (from short to long term). Once the method of complexity management is specified, the system can be modeled adequately for the final implementation. In the following sections, the steps visualized in Fig. 6.1 are described in detail. Special attention is paid to the task of information acquisition, as this represents a difficult and laborious task, which heavily impacts the entire process outcome.

Complexity appears in many systems and contexts, which explains the variety of viewpoints, definitions and methods for handling complexity [1]. Examples clarifying the steps of this guideline have been chosen from the field of structural complexity. The reason therefore is that this kind of complexity and associated models and methods are applicable and widely used in many engineering fields. Structural complexity management [2] represents one specific view on complexity and has been applied on many challenges concerning technical, process-based or organizational networks [3, 4]. From this perspective, system elements and their dependencies form the core of a complex system. Senge and

Stermann describe this definition of complexity as detailed complexity (in contrast to dynamic complexity) [5, 6]. In structural complexity management not only the number of system elements and dependencies, but also their resulting constellations serve as input for analyses [7, 8]. Several system characteristics and the behavior can be conducted from these analyses [4]. The abstract character of the guideline in Fig. 6.1, however allows for the application of any kind of complexity modeling.

6.1   System Definition

Dealing with a complex engineering challenge often implies that the problem has not been understood completely. Thus, an instant system analysis seems not to be helpful on such a basis of incomplete information. Consequently, the initial task needs to be defining the system, as it helps clarifying the scope of the considered problem, system boundaries, considered system components and general objectives.

The initial question of the system definition should be about the origin of complexity. It is important to mention that consequences of complexity and its initial source can be located in different places. For example, one can observe extreme difficulties in managing the assembly process of product components in order to build a variety of products. However, the observed complexity (complexity impact) can be the result of a non-optimal portfolio of product components. That would mean that the origin of complexity is located in the area of product components, while negative impact can be observed in the process areas. If one would not question the origin of complexity, one would likely tend to create a process model and take measures of assembly optimization, and the origin of complexity would not even be part of the model.

A subsequent second question should be for the potential impact of complexity in the considered system. This is important as the pure existence of complexity is not a sufficient reason to take measures. For example, while complex effects happen during the combustion process in a car engine, this does not impact a person while driving. Even if the complex processes are located inside the car, they are not part of and nor linked to the driver's system. Consequently, this complexity does not impact the driver's system.

In the case when complexity has impact to a system in question, this leads directly to a subsequent question about the potential harmfulness of complexity. In many systems it is easy to identify a kind of complexity. For example, more detailed modeling increases the number of system elements and their interconnections—resulting in more structural complexity. However, the pure existence of complexity in a system does not necessarily require any countermeasures, if the effects resulting from this complexity are not harmful. A harmful effect could for example be increased uncertainty about component impacts within a product, which then leads to numerous unexpected and cost-intensive constructive changes to product components.

If system complexity with harmful, undesired impact could be identified, the next question should be about the objective of a complexity analysis. Reduction or avoidance

Table 6.1 Questions for clarifying the system definition

#	Question	Examples
1	Where does complexity in the system emerge from?	Dependencies between components or process steps
2	Which effects result from system complexity?	Unexpected change propagations, unexpected process delays
3	Which effects are harmful?	Expensive and time-demanding adaptations
4	What is the objective of complexity analysis? Which results are expected?	Possibilities for the improved development of a product

of complexity per se cannot be an objective. Benefits can only be achieved by decreasing negative or harmful impacts, which are the consequence of existing complexity. Defining the desired project results helps to specify clear objectives and makes project progress and success measurable. Table 6.1 summarizes the introductory questions for clarifying the system definition. Guided by these questions, the general system elements and dependency types, levels of detail, system states and boundaries should be specified.

6.2   Identify the Type of Complexity

With a definition of the considered complex system on hand, the next step is to identify the type of observed complexity. This step partly overlaps with the system definition, which already specifies the source area of complexity (and distinguishes it from the area of complexity impact). Lindemann et al. distinguish four main areas of complexity in engineering design: market, product, process and organizational complexity [2]. This classification can be helpful, as for each area of complexity different approaches of management have been developed and established, e.g. product modularization [9] in the product area or process streamlining in the process area [10]. Depending on the specific challenge, a different set of categories can be helpful too. For example, Ashkenas looks at complexity from an organizational perspective and defines managerial behavior, process evolution, product and service proliferation and structural mitosis as the four sources of complexity in organizations [11].

In addition to this classification, complexity can be described by characterizing it as useless or useful. An easy example explains this: If an enterprise produces customized production plants, significant complexity could emerge from the large number of customer requirements, functions and components. And this quantity can create complexity, for example if technical adaptations are required and component interdependencies result in undesired change impact. Also the large number of components can cause significant process complexity, for example when managing customized product assembly. If in this example, the step of system definition resulted in identifying the source of complexity as being the numerous requirements, functions and components, then complexity could be reduced by standardizing the so far customized plants. In fact, standardization is a common strategy for complexity reduction. However, while this strategy would reduce the complexity, in this use case it would definitely also influence the company's market offer negatively.

In this simplified example useful complexity has been reduced, resulting in a negative effect to the company. In fact, in many cases complexity contributes to a competitive advantage, if this complexity can be handled. Then complexity is useful, and even increasing complexity could be beneficial—providing that this complexity can still be managed. A common example for such an increase of complexity would be the further enlargement of a product portfolio. With such a step companies intend to cover more market niches and increase the company's success.

Anderson describes several cases for useful complexity in his book "The long tail" [12]. He explains that enlarging the product portfolio even with low-selling products can be beneficial in today's economy. As positive examples he quotes Amazon with its tremendous amount of offered products. However, it has to be mentioned that Anderson describes the business models of companies offering mutually independent products via Internet stores. If one product is added to the portfolio the complexity does not increase significantly, as products do not affect each other (logistic considerations are neglected in this example). However, integrating an additional variant of a component in a car could have highly complex effects, as this would affect the whole system due to the large number of dependencies between components. This is the reason why Anderson's concept of "selling less of more" so far works for non-interconnected product portfolios only. But it shows that the potential of managing larger amounts of complexity could result in increased market strength. In other words, being able to manage useful complexity and therefore being able to increase it can create competitive advantages.

After answering the questions in the system definition step and considerations about the usefulness of the observed complexity in question, one should have a solid basis for selecting an appropriate complexity management strategy.

6.3   Strategies and Associated Methods for Handling Complexity

In many situations observed complexity appears as a negative system characteristic, which people dislike being confronted with. Thus people tend to implicitly chose the strategy of reducing complexity. But the explicit selection of an appropriate strategy for approaching complexity challenges is important—and therefore represents a step in the complexity management framework introduced here.

Available resources and the time frame for required results are the main criteria for selecting an appropriate strategy. Figure 6.2 links the three strategies with their time scope (adapted from [13]). In general, strategies for handling complexity can be classified into three groups: avoidance, reduction and management.

Reduction of complexity can be applied as a short-term measure for minimizing already existing complexity [14]. If, for example, the offering of a complex product spectrum with

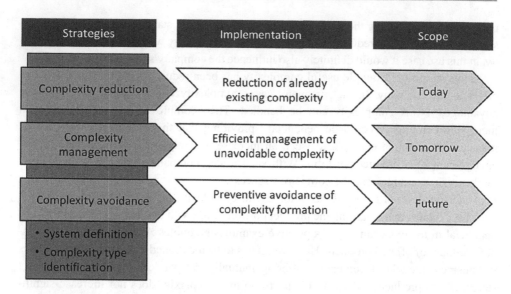

Fig. 6.2 Approaches towards handling complexity (adapted from [13])

extensive component variety results in process and production failures, reducing the product variety instantly reduces complexity. However, one must be aware of the fact that such a reduction is not going to the root of the problem; instead it only attenuates the consequences. For this reason a strategy of complexity reduction should be accompanied by more comprehensive complexity handling measures once the urgent need is satisfied. For the example of the complex product portfolio, the root cause of the large variety of offered products could be a sales concept that accepts too many customization requests. Thus, if the offered product portfolio is only reduced once and no further measures are taken, complexity would increase again over time.

As mentioned above, complexity can be viewed as a threatening experience. So it is understandable that complexity reduction is often seen as a generally useful measure. However, also beneficial types of complexity exist, because it leads to a competitive advantage. If, for example, a large product portfolio offer is decisive for market success, then the reduction of associated complexity would result in a negative impact, i.e. decrease the competitiveness of the company [2]. Therefore it is important to identify the type of complexity (useful or useless) before selecting a strategy of complexity management.

Management (controlling) of complexity is a strategy for handling unavoidable internal complexity, which arose through external sources (and therefore is unavoidable) [13, 14]. This strategy includes measures for complexity handling in development work and process design. An example could be the implementation of a configuration tool that supports the assembly of customized products, or a knowledge management concept that supports development work by increasing the reuse of existing solutions. As the implementation of such measures requires significant effort, this complexity management is a strategy with at least mid-term time scope. The term unavoidable may create the impression that this type of complexity is undesired and useless in terms of the company's success. Therefore it needs to be mentioned that when complexity is useful in terms of a company's competitiveness, this complexity should also not be reduced or avoided—and therefore it is unavoidable.

Complexity avoidance is a strategy oriented towards the future [13]. It represents the consequent focus on the root causes of complexity emergence. This strategy is not useful as an ad hoc approach, as it is aiming at preventive avoidance and therefore has no effect on the status quo. In addition, measures of complexity avoidance typically require a significant effort in implementation and therefore aim at a long-term scope. In terms of the product portfolio example, possible measures could be the implementation of entrance barriers for the creation of new product components. For example, if the sales department requests the development of a new component, then one decision criterion could be a minimum purchase quantity. A certain threshold of purchase orders then needs to be met before development work is initiated. This could avoid the uncontrolled inflation of the product portfolio by component variants that only get ordered once. This and similar measures of complexity avoidance often require significant changes of business processes and even a company's culture.

Especially if companies do not have complexity management established permanently into their processes, they notice the need when complex problems negatively influence their daily business. In this situation, measures of complexity reduction and control have to be applied first, but should be accompanied by introducing complexity management for future success. The fundamental changes required for many approaches of complexity avoidance (as well as comprehensive complexity management/control approaches) need forward-looking planning and cannot be implemented as ad hoc solutions in crisis situations.

The necessity of treating useful and useless complexity differently has been addressed above. In addition, it needs to be mentioned that successful complexity management can turn initially harmful complexity into useful complexity. As long as a huge product portfolio cannot be controlled, process and production failures can be harmful to the company. And a company could be tempted to reduce product components and variants—even if a broad market offer would confer competitive advantages. The better the means of complexity management, the more complexity a company can support and apply for its success. Consequentially, the increase of useful complexity should also be named as viable strategy for handling complexity, assuming that the adequate management tools are on hand.

It should be noticed that useful complexity does not only have to be located in the product domain (as described in the example of a complex product portfolio), but could also emerge from complex processes that serve the customer demands. Examples could be advanced possibilities of product configuration and fast delivery by means of optimized business processes.

Also the relocation of complexity can be observed in practice: In the early years of mobile communication, Nokia became one of the dominant players in the market with an impressively large product portfolio [15]. Managing such a portfolio with short product life cycles is definitely challenging. But the variety guaranteed Nokia's strong position in the market. Then with the advent smartphones, Apple Inc. entered the market with its iPhone product in 2007. Even though it was only a single product variant, in 2008 customization became possible by installing a personal selection of (third-party) apps. Building up an ecosystem of software applications, assuring their functionality and safety and managing payment services created new challenges (and opportunities). So the useful, market-relevant complexity associated with a large product portfolio was shifted from hardware to software products.

Another strategy of complexity management is the adequate representation of the complex issues. As already introduced, complexity means the lack of transparency. If a complex system, its elements and dependencies can be visualized, then the resulting easier access to the system increases system understanding and possibilities of system interaction. Especially if several people are required for solving the problem, a suitable visualization can facilitate the discussions and prevent misunderstandings.

Methods provide procedures that support systematic problem solving. Many methods aim at managing complexity, with different approaches. Selecting an appropriate method for solving a complex problem can be challenging, because the ambiguity of complexity hinders clear determination of a method. In other words, it is difficult to select the right tool without knowing what to fix. Application of the guideline shown in Fig. 6.1 shall help minimize the uncertainty when selecting an appropriate method. This systematic approach also clarifies the risk of selecting a method without sufficient previous analysis of the complex problem. In complex, non-transparent situations people tend to stick to methods they are familiar with—especially if time pressure is an issue; however, this does not imply that the method is suitable for the specific challenge. The following sections describe some methods for handling complexity depending on the selected strategy.

6.3.1   Create Transparency by System Views

As a lack of transparency is a main characteristic of complex problems, it seems likely that comprehensible system views can facilitate solving them. Especially when system knowledge is shared by several experts, a comprehensive system view can focus the effort on a common target. Methods for creating transparent system views aim at element and dependency acquisition, preparation of system views and the enablement of interactions of experts with those views.

A large variety of representations exist for modeling complex problems. For the creation of transparent system views highly sophisticated approaches are less useful, as they do not provide easy access for many people, especially if they come with different knowledge backgrounds.

Graph notations have especially become very popular for easy-to-understand representations of system elements and their interdependencies. Reasons therefore are their simplicity and wide-ranging applicability. Computer-supported tools use different mechanisms for element alignment (e.g. force-directed graphs) and additional information representation (e.g. size, form or color of edges and nodes) for creating transparent system views, which often even allow for dynamic interactions [16, 17]. The requirements of big data analyses and progress in computational power bring up continuously improving graph representations with more and more functionalities [18]. But even if technical possibilities allow for representing an enormous amount of system details, it has to be taken into account that the human visual capacity is limited.

One solution to this limitation is the application of specific system views, which allow extracting relevant information from comprehensive system models in order to provide a manageable amount of information to a user. Transparency improves from highlighting only specific aspects of a system. This act of focusing on parts of the system can be done in two different ways:

The first possibility is to isolate specific parts of a complex system for representation. This means that all system parts that are not required for understanding the current question get neglected in this representation. If, for example, the collaboration within a large development department should be investigated, the visualization model could focus on the exchange of documents between employees. A graph representation would then allow the identification of centrally located employees as well as closely related employee groups. When this approach is selected it has to be assured that the highlighted system view permits one to draw meaningful conclusions. And it is important that the extraction of partial aspects does not provoke any misinterpretation, as they can easily appear: in the above example of an organizational structure the collaboration between employees could result not only from the document-based information flow, but also from informal communication, meeting structures, commonly developed components or the work on shared projects. If only one aspect of communication is extracted, accurate conclusions on the entire communication cannot be drawn.

The second possibility of creating transparency by highlighting specific aspects is to aggregate interdependencies into a system view. For example, dependencies like document exchange and component responsibility in a design department could be superimposed in order to create a general "dependency view". While the specific reason for a dependency gets lost in such an aggregation, the density of represented information decreases while not neglecting any information. A detailed approach towards the creation of specific system views is described in the book Structural Complexity Management [2]. This approach is based on matrices as the basic tool for information acquisition and aggregation, whereas the visualization of resulting system structures can be done in matrix and graph format. Such system views provide better transparency for users than interacting with an unmanageable number of elements and dependencies in an unrestrained modeling of a complex system. In general, the creation of systems views should be only as detailed as necessary. It is not important which information is available, but which information is necessary to answer the question.

Lindemann et al. create system structure models based on two fundamental types of matrices, domain mapping matrices (DMMs) and design structure matrices (DSMs) [2]. A DMM is a matrix linking elements belonging to two different groups (domains), where typically each axis contains one group (domain) of elements. DSMs link elements belonging to the same group (domain); thus each element is shown on both axes of the matrix (the element order is identical on both axes). Eppinger and Browning provide details about the theoretical and practical application of these matrices [19].

Interestingly, all kinds of complex system structures can be built up based on the two matrix types DMM and DSM as their basic elements. Figure 6.3 shows the exemplary modeling scheme of dependencies between processes and documents. This scheme indicates the fact that executed processes deliver information to documents and that other processes require this information stored in documents. Both dependency types can be modeled using DMMs, as they link two different types of elements. A simple mathematical operation (matrix multiplication) then allows one to derive an aggregated network of processes indicating the information flow between them [4]. This aggregation suppresses the visualization of documents and integrates their links to processes into newly modeled links between process elements. While the fundamental DMMs are easy to acquire (the required information is often on hand), the aggregated process network provides an easy-to-understand, streamlined system view, which makes the process network transparent.

Figure 6.4 shows a more comprehensive modeling scheme of a system with product components that influence each other, people interacting with product components and

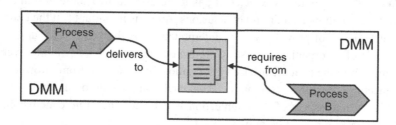

Fig. 6.3 Linking processes to documents by two DMMs

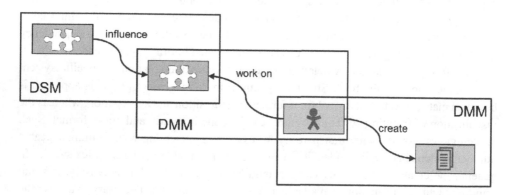

Fig. 6.4 Linking components, people and documents by DSM and DMM

documents created by people. This system can be decomposed into two DMMs and a DSM, as it is indicated by the rectangles in the figure. One aggregated view of this system could e.g. show mutual impact between people based on their work at (mutually linked) components and their common work on documents. In this case, components and documents would be suppressed and integrated into newly created links between people. Visualized as a transparent "impact map", this system view could be used for several decision-making processes.

6.3.2 Avoid or Reduce Complexity

In case of existing useless complexity, a strategy of complexity reduction can be applied. Reducing complexity is well-established for application on useless complexity of a system. Useless complexity means variants that increase costs and effort without adequate value in the market. Several methods in the engineering design field are applicable for implementing this strategy.

Especially variant management needs to be mentioned due to its significant relevance in the field. When aiming at the reduction of complexity, methods of variant management are applied for identifying product and component variants with undesired characteristics, e.g. low sales figures. If the removal of these identified variants does not impact other, desirable variants, the product portfolio can be revised accordingly. Identifying if and how components are interdependent in a large product portfolio represents a challenge, typically tackled by network analyses and intensive use of rule sets. Several authors provide comprehensive introductions to variant management and method sets for practical application [9, 20].

Obviously, reducing useful complexity cannot be a reasonable approach. However, many systems contain both useless and useful complexity, and it is important to reduce only useless complexity when applying this strategy. As well, one needs to keep in mind that so far useless complexity can be turned into being useful complexity if it can be controlled. Established evaluation methods can be applied for identifying useful and useless complexity. For example, a Pareto analysis facilitates the classification of products in terms of their profit generation and hence can separate useless from useful elements. Within a product portfolio, all products below a certain sales figure or above a certain production effort could be identified with this method. Value analyses or target costing approaches provide similar possibilities in terms of specifying the contribution of system elements to useful or useless complexity. In all these cases the selection of an appropriate evaluation criterion is of major importance.

Otherwise, product variants that add significant value (e.g. market strength) to the company at affordable costs represent the useful part of complexity. Schuh and Schwenk do not use the wording of useful and useless complexity, but their approach of optimizing the number of variants in a product portfolio means finding a subset with only useful complexity. It needs to be mentioned that their approach is not about reduction only.

Finding the optimal number of product variants can also mean extending complexity to fulfill untapped market potential [9].

Anderson explains that many markets ask for increasing product variety [12]. As the optimal number of product variants for an enterprise is determined by a tradeoff between the market value of variants and their costs (or complexity) to the enterprise, methods that decrease these costs can in return allow the increase of variants (useful complexity). One starting point can be reducing complexity in processes. Several authors describe methods for minimizing iterations, e.g. based on matrix approaches [7, 21].

Avoidance of complexity is a long-term strategy with the objective of not letting complexity emerge—and so no need arises to reduce it later on. Baldwin and Clark describe for product complexity that a method of decoupling product modules by interface design can help in avoiding complexity [22]. Such decoupling breaks up interdependencies between the modules and makes them mutually independent. Adaptations to one module then do not impact other modules and therefore avoids complexity resulting from change propagation.

For avoiding product portfolio complexity, defining entry barriers for the adoption of new product or component variants can be helpful. Companies, whose business models are not based on continuous product customization can efficiently avoid the rise of new variants. The selection of suitable evaluation criteria is of major importance. Wrongly chosen criteria can either be inefficient (so that undesired variants still emerge) or block meaningful variants from being introduced. That would mean that wrong criteria would hinder useful complexity from being built up.

6.3.3   Manage and Control Complexity

Here, managing complexity is understood as controlling it. Methods for controlling complexity primarily aim at the lack of system transparency as a main reason for negative impact from complexity. In highly networked systems then, decision making cannot be executed based on simple cause-and-effect chains, as they are either not accessible or, when extracted from the system, represent an incorrect simplification (neglect of side effects). The number of elements and dependencies in a complex system prevents one from acquiring as well as representing it completely. However, this is not even necessary. And it is important to notice that our daily decisions are hardly ever based on complete problem descriptions with all information on hand. In fact, successful decision-making is based on models, which contain only the relevant information. Thus, the challenge of managing complexity is not to acquire all the information, but to identify the relevant parts and to make them accessible in a suitable, comprehensible model. For this reason the early steps of system definition and identification of complexity types are of major importance.

Wildemann describes managing complexity as a strategy required for dealing with unavoidable complexity [13]. This can be misleading, as unavoidable still means that this complexity is undesired, but seems not to be removable from the system. However, useful complexity is desired and contributes to an enterprise's market success. The better that useful complexity can be controlled, the more can be included in the system.

This can be explained with the example of an enterprise offering components for production automation. If the business model of the enterprise requires providing a multitude of components with high variability (e.g. different power ratings) and with the possibility to assemble different components into systems, then methods and tools for improving the configuration can be helpful to be implemented. Based on configuration rules, these methods and tools allow managing the products, variants and their possible combinations even on a large scale. Approaches on such internal configuration guarantee the development and integration of components and variants, which are compatible with the company's existing portfolio and extend it meaningfully.

Methods and tools for managing external configuration facilitate managing complexity at the interface to the customer. Customers can apply configuration managers for compiling and selecting customized solutions from a large, unclear spectrum of choice. Many configuration approaches try to make selections easier for the customer by letting him specify his needs instead of the technical details. The configuration tools then link the user requirements with the best matching product configuration. Felfernig et al. provide a profound introduction to the history and state of the art in configuration techniques and tools [23].

Schuh and Schwenk describe methods for matching internal complexity (number of component variants to be managed by the company) with external complexity (product variants visible to the customer) [9]. The objective is to realize a large external (useful) complexity with a small-enough-to-manage internal complexity. The better a company can manage this internal complexity, the more it can handle and consequentially extend the external market offer.

Other methods of managing and controlling complexity target the typical lack of understanding that goes along with complex systems. Daenzer and Huber describe checklists for complex system structures as a systems engineering approach, and Maurer explains them in the context of structural complexity management [4, 24]. An input checklist focuses on a specific system element (e.g. product component, process step or organizational resource) and indicates other elements that impact it [24]. This provides the possibility to monitor potential disruptive elements and foresee and control change propagation before it happens in an unregulated way. Input checklists can be applied beneficially on system elements that should not change, e.g. because of high effort, cost or safety issues related to such changes.

Feed-forward analyses are similar to input checklists, but do not serve for monitoring potential impactors but potentially affected system elements—emanating from one specific system element [4, 25]. Such an analysis allows estimating the consequences of adapting a specific system element by investigating the impact of this change to its surroundings. This perspective can be helpful when applied to those system elements, which are likely to be changed in the future (e.g. because customization requests are attached to these elements). If feed-forward analyses are already prepared for selected elements, once the specification of a change request is on hand, consequences can be systematically discussed and managed [4].

Trace-back analyses are meant for finding the element that initiates an observed system behavior or effect. Starting with the observed element, the system structure gets traced backwards until the initiating element is found [4, 25]. Depth-first search, breadth-first search, branch & bound and branch & cut are some methodical approaches to organize and optimize the search. Visual support, e.g. by graph representations, can also be very effective.

Daenzer and Huber introduce a simple but powerful influence portfolio, which allows one to identify the general embedding and parts of the behavior of selected system components [24]. For each system element, the outgoing (active) dependencies to other system elements are indicated on the horizontal axis. On the vertical axis, the incoming (passive) dependencies from other system elements are noted. The relative location of each element in the diagram allows one to deduce the basic behavior of an element.

Figure 6.5 shows an influence portfolio for the main components of an automotive airbag system [2]. The diagram is divided into four areas. In the lower-right area, elements with mainly outgoing and only few incoming dependencies are located. Those are active

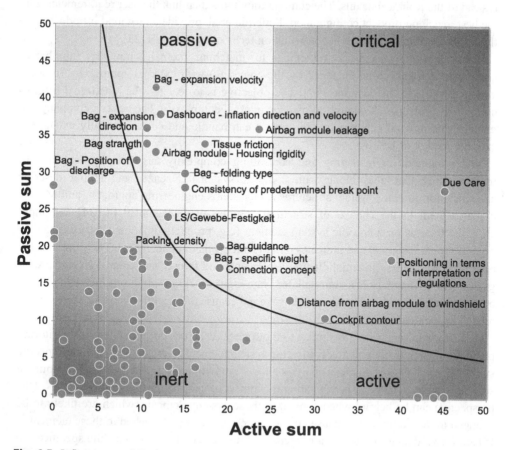

Fig. 6.5 Influence portfolio for an airbag system [2]

elements, which can influence the whole system via their many connections, but do not get influenced much from other elements. Elements located in the upper left corner behave contrary and are called passive elements. These elements possess many incoming and only few outgoing dependencies. Elements in the lower left corner are call inert elements, as they generally do not possess many connections to the system and therefore do not influence and are not influenced by the system very much. Elements located in the upper right corner represent the so-called critical elements and they possess the most dependencies within the system, having incoming as well as outgoing dependencies. Because of the many dependencies it is very likely that these critical elements are involved in any kind of changes to the system.

The criticality of an element is computed by multiplying the number of outgoing (active) dependencies with the number of incoming (passive) dependencies. The curve shown in Fig. 6.5 represents a line with constant criticality. When applying the impact diagram, typically such a line is selected as a threshold for identifying these elements, which need to be more closely investigated or monitored because of their high criticality. Of course, setting a value for this line is a subjective decision. In the project that created the diagram shown in Fig. 6.5, all elements with a higher criticality than indicated by the line became listed on a checklist, which had to be discussed for possible problems whenever a significant change anywhere in the airbag system was planned. As well, planned changes to any one of these highly critical elements had to be approved by the project management.

Methods similar to impact analyses, feed-forward and trace-back analyses are known by other names, e.g. cause-and-effect analyses. All these methods are based on modeling the complex system with nodes and edges in network form. These methods are meant for systematically managing changes to the system by increasing transparency and reducing uncertainty even when working with systems consisting of a large number of elements and dependencies.

Product modularization is often applied for managing and controlling complexity in large product portfolios [9]. From a system structure perspective, modularization means bundling highly interlinked product parts into modules and combining these modules by standardized interfaces, which means fewer and easier to control dependencies. In the same way as modularization, platforms or building blocks as well as differential and integral product design represent useful methods of managing useful complexity [4, 22].

Creating transparent system views, as explained in Sect. 6.3.1, is also a powerful technique to support controlling complexity. It allows people to interact with a complex system, detect anomalies, experiment with if-then-scenarios and discuss possible measures. This can be explained using the visualization of a student race car in Fig. 6.6 (adapted from [4]). The force-directed graph shows 27 main components with their interdependencies. The alignment of elements intuitively uncovers some basic system characteristics. These can be interpreted and applied for managing system interactions.

One central element (frame) connects three main subsets. The central embedding and the many dependencies of this element indicate its critical importance to the system. Three elements (gear shift, data logging and wiring harness) are isolated from the other parts of

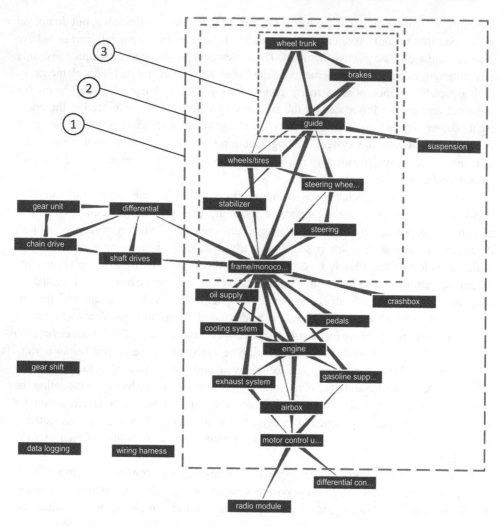

Fig. 6.6 Transparent system structure for facilitating complexity management (adapted from [4])

the system; also four elements (suspension, crashbox, differential control and radio module) represent leaves, which are only connected to one other element of the system. This minimal connection to the system makes it easy to control impact to and from these elements.

Basic tool-enabled graph analysis can facilitate further interpretation of the structure. All elements located within the frame indicated with 1 make part of the same strongly connected component. This constellation is defined as a set of elements, which are (indirectly) connected to each other by at least one path. From a control and management perspective, it is relevant that strongly connected components encapsulate feedback loops. So, cyclic effects as described by system dynamics models can only occur within this constellation.

In densely connected system structures, strongly connected components can become very large. And further subdivision of the structure can be helpful. The frame indicated with 2 in Fig. 6.6 contains elements forming a bi-connected component, also called a block. In this constellation all elements are mutually (at least indirectly) connected by two separate paths. Thus the integral character of a block is significantly higher than in a strongly connected component. Because of the two paths between each element, for example possible change impact cannot be avoided by simple, single measures. Engineering work at physical components forming a block structure require intensive organizational collaboration. An intensification of the block is indicated by 3 in Fig. 6.6. This constellation is called a complete cluster or clique; all elements are directly connected to each other and adapting any one of them always requires consideration of all other elements.

In general, interpretation of structure visualizations can support the control and management of complex systems. In this context it is worth mentioning that structural interpretations make part of peoples' daily life. We identify bottlenecks in processes or resource allocations and so put an interpretation to an articulation node, as it is called in graph theory. And when looking at an organization we talk about a key person, when in a graph depiction this person would be centrally embedded to the structure with many interdependencies to other parts of the system. In an impact diagram, such a person would be characterized as highly critical to the system.

6.4 System Modeling: The Challenge of Information Acquisition

Representations and analyses of complex system structures often use matrix approaches. Lindemann et al. describe a generic five-step approach on structural complexity management using matrices, which are complemented by graph representations [2]. Especially the task of information acquisition is challenging, because "the strength of the result and interpretation depend greatly on the reliability and validity of the data and the assumptions" [26]. Furthermore, information acquisition can be extremely resource-demanding, which can lead to problems of fatigue or training effects [2, 27].

Several requirements have to be considered in order to model the relevant system information. With those in mind, a suitable method of information acquisition has to be selected and applied. This application can be impeded by several possible errors, which can yield inaccurate models. Therefore, requirements, methods and possible errors will be described in the next sections.

6.4.1 Requirements for Information Acquisition

Information acquisition represents an important task in any approach on structural complexity management. In this context, several authors mention requirements, which have to be fulfilled for reaching high-quality system structures. In the following paragraphs these requirements are aggregated and assessed concerning the need for methodical support.

For information acquisition using a design structure matrix (DSM), Dong explains the importance of adequate quality and completeness of information acquisition with its direct impact to subsequent tasks like analysis or interpretation [28]. Whereas the quality of input information and analysis output correlates positively, a negative correlation can be stated between increasing input quality and economization of resources. "Since the DSM is a tool to analyze the design project and to seek improvement, it is important that the data is accurate. When necessary one has to trade the speed of data collection with the quality of the data" [28].

The quality and completeness of structural information is difficult to measure. Dong refers to the interaction density ratio as an indicator for system model quality based on the number of system elements and dependencies [28]. If a system structure is represented in a DSM, this ratio compares the total number of off-diagonal marks with the total number of rows in a DSM [29]. Dong hypothesizes that an interaction density ratio of 6 indicates that enough information about a system has been acquired [29]. Maurer proposes the use of structural constellations (e.g. articulation nodes or star-shaped structures) for plausibility checks during an acquisition process [4]. These plausibility checks can also be based on the existence of direct and indirect dependencies between elements, and make criteria for deciding about the need for iterative information acquisition [30].

Bartolomei mentions that the reason for a lack of quality of a system structure can result from focusing on technical knowledge only [26]: "The engineering domain is methodologically ill-equipped to describe and represent components beyond the technical domain, such as describing the factors that influenced design decisions, mapping social interactions, and understanding systems processes. This is problematic for systems-level modelling frameworks as they are designed to represent knowledge that spans the social and technical domains" [26].

6.4.1.1 Adequate Level of Detail

Modelling system elements and dependencies too abstractly will only provide trivial insight; too detailed and the models can hardly be acquired within a given time and budget. For this reason an adequate level of detail is of major importance for successfully conducting information acquisition. Bartolomei criticizes a model describing a jet turbine with only 60 elements [26]. The author argues that the model of a highly complex system cannot provide significant benefit if it gets simplified too much. In contrast to this, Browning discusses resource problems arising from too extensive system structures [3]. He proposes an aggregation of system elements, i.e. modelling the system in less detail when the system size exceeds manageable amounts. Dong aggregates elements by defining three general segments: social, technical and natural [29]. She declares that dependencies within and between all three subsystems have to be acquired for obtaining a complete model. In hierarchical order, Dong defines component-level knowledge as the detail-level knowledge, which can be acquired from experts. However, system-level knowledge is required for describing complex systems and needs to be assembled from the component-level knowledge of several experts [29].

Biedermann et al. also mention the importance of an adequate level of detail and suggest top-down modelling of system structures [31]. In this approach, dependencies are only acquired between top-level elements in the beginning. Next, possible dependencies at the detail level are investigated. Such dependencies can only exist if their superior (high-level) elements have already been linked. This means that large numbers of possible dependencies at the detail level can be excluded from further consideration if no high-level dependency has been acquired first. With this procedure Biedermann et al. try to enable modelling at a detail-level while keeping required resources at a minimum.

Not only system elements, but also dependency descriptions define a system's level of detail. Rowles mentions that the application of dependency weighting (instead of modelling the existence/non-existence only) can be unfavourable if interviewed experts cannot provide reliable information [32]. This means that the applied level of detail and the available system information have to match.

6.4.1.2 Accessibility, Traceability and Extensibility

Lindemann et al. describe that information acquisition processes can be highly iterative [2]. This is why acquired system elements and dependencies need to be accessible at any time. Dong indicates the DSM as being favourable, as "DSM acts like a browser that provides user directions to use the existing information database" [28]. Reasons for the declaration of elements and dependencies should be documented, as this can help in retracing decisions made in earlier acquisition processes. Dong highlights the importance of up-to-date documentation of all system elements in case of later system extension [29].

Especially for the modelling of large structures, the application of databases can facilitate the interaction with the model. For example, Ahmadi et al. describes the acquisition of a product design process by using a database [33]. Accessibility, traceability and extensibility are directly connected with two further requirements: the need for a systematic procedure and appropriate methods and tools, which will be introduced next.

6.4.1.3 Systematic Procedure

A systematic procedure of acquisition is a precondition for later accessibility, traceability and extensibility of system structures. As well, adequate quality and completeness of structures only seem to be achievable with such a procedure. Eppinger describes a procedure based on the compilation of a DSM [34]. He explains that system elements have to be acquired first, followed by the dependencies. Eppinger and Salminen mention that knowledge owners should be interrogated about their required input: "It's important to focus on input rather than output because we have found that managers, engineers, and other product-development professionals are more accurate in identifying what they need to know than in describing what others need to know" [35].

Black et al. propose separating the acquisition of system elements from the subsequent acquisition of dependencies [36]. This procedure allows one to only ask knowledge owners about dependencies between elements under their own responsibility. As structure acquisition is time-consuming, this way every expert only gets burdened with the minimum of dependencies. Another approach is the separate acquisition of organizational and product dependencies. Both systems can then be compared to each other and the results may lead to iterative structure acquisition [32]. Pimmler and Eppinger propose an interrogation scheme, where information provided by knowledge owners is documented together with initially formulated estimations [37].

Some authors see the need for more detailed acquisition processes [28, 38]. Avnet describes a four-step procedure on information acquisition, which includes an intensive mix of acquisition techniques [38]. Observations of team meetings, surveys and interviews get combined and applied to a verification process by a team leader. Dong introduces a seven-step approach, which separates the collection of system elements and dependencies from subsequent system documentation [28]. Knowledge owners are interrogated by the system modeller, but they are not involved in formulating the system dependencies in the model afterwards. Bartolomei proposes a procedure called "qualitative knowledge construction", which consists of eight sequential steps [26]. This procedure applies "thick data", as mentioned in the grounded theory [39]. Sabbaghian et al. apply software for assuring the systematic execution of information acquisition [40]. The authors criticize other methods of information acquisition (i.e. interviews and meeting participation) as being too resource-demanding and too difficult to coordinate.

The examination of requirements for structure acquisition shows that high-quality information has to be generated at the right level of detail. Information needs to be accessible at any time and should be acquired in a systematic process. Methods for fulfilling these requirements will be described in the next section.

6.4.2   Methods of Information Acquisition

A detailed survey and classification of applied acquisition methods allows us to conclude the topic described above of fulfilling acquisition requirements. Almost 100 publications could be identified which contain structure acquisition by expert interrogation. In most of these publications, the act of conducting interrogation was only mentioned without any further information about applied processes or methods. Only a few contributions describe methods, advantages and disadvantages in detail. These publications have been examined and six basic methods of information acquisition can be extracted.

The investigation of scientific publications in terms of the origin of presented system structures shows that information acquisition is only considered sufficiently in a few cases. This means that the correctness of many analyses, interpretations or optimizations could be doubted, because the quality of applied structures is often unclear.

From over 500 investigated contributions to journals and conferences (all dealing with system structures) almost two-fifths do not describe any conducted information acquisition. Of course, one cannot conclude that in these projects information acquisition was not executed. But obviously this step and its documentation was not of primary importance in these contributions. Now, without clear statements about the input information and their acquisition, it is difficult to evaluate the results. Another one-fifth of the investigated contributions avoided the effort of information acquisition by either applying system structures provided by other publications or creating exemplary system structures themselves. In another one-fifth of the surveyed publications, structural information is extracted from data sets, documents or models. Here the authors make use of a highly effective form of information acquisition. Especially if the import of data can be automated, large-scale structures can be obtained without extensive acquisition effort.

If the acquisition of system structures requires the interrogation of knowledge owners, then methods like workshops or interviews are required. Such methods were applied in one-fifth of the publications, and then the methods in these publications were examined in greater detail. Unfortunately, most authors who mention acquisition by interviews or workshops do not provide specifications of the acquired systems; therefore it is hardly possible to assess the effectiveness of approaches described.

Only seven publications mention the size of the system that had been acquired. Here, the numbers of considered system elements varies from 18 [41] to 600 elements [42]. Only six authors specify the duration of the acquisition process, which varies from a few hours [43] to several months [44]. Details about the number of knowledge owners involved could only be identified in eight contributions. Numbers vary from 6 [43] to 70 people [10].

In general, acquisition processes are rarely described. And it seems reasonable to suppose that the lack of descriptions goes along with insufficient consideration of information acquisition. As the acquisition of a system structure represent the basic input for subsequent modelling, analysis and interpretation, available methods and challenging barriers have to be known. Methods can be classified into six groups: analog survey, digital survey, documentation, observations, interview and estimations. In the following paragraphs these methods will be detailed. Table 6.2 sums up advantages and disadvantages of each method.

6.4.2.1 Analog and Digital Surveys

An analog survey means to investigate documents which have been filled out by system experts. Multiple-choice questions are commonly applied, because they are easy to process for the interviewee and the evaluation can be automated. An obvious advantage of analog surveys is time efficiency, as many experts can be interrogated without much effort [28]. Dong also mentions possibilities of statistical analysis as advantageous. Browning acquires the development process of an aerospace company using an analog survey [45]. He observed that experts have different understandings of the questions asked, which can influence the resulting quality negatively. In addition, he mentions that incomplete responses to the questionnaire can be problematic. Rowles describes analog surveys for acquiring the system model of a jet engine and the associated organization [32]. Similar to Browning, he describes the risk of a low response rate as disadvantageous. Avnet let the participants of design sessions fill out analog survey forms after each session [38]. He, as

Table 6.2 Methods for the acquisition of system structures

Methods	Advantages	Disadvantages
Analog survey	Time-saving [28] Possibility of statistical analysis [28]	Incomplete response to questionnaires [45] Different understanding of concepts [45] Random assessment of quantifiers [45] Multiple-choice constricts creativity and solution space [38] Information loss by constricted responses [28] No justification of responses [28] Specification of target-state instead of as-is-state [28] A priori problem [26] Low response rate [32]
Digital survey	Possibility to process large quantity of data [40] Requirements on resources minimized [40] Possibility to model dynamics over time [40]	Unclear terminology leads to rework [40] Responsible interaction of several people required [40]
Documentation	Low effort, automatable [2]	Outdated data [28, 45] Unrealistic data [28] One-sided representation [28]
Observations	Possibility of fast conflict resolution [28] Understanding of situations facilitated [38] Interpretations of surveys or interviews [38]	Potential information loss because of hierarchy and social pressure [38] Subjectivity of observations [38]
Interview	Identification of direct dependencies [28] Identification and resolution of disagreement [28]	Danger of intentional concealment of the real process [26] Time-consuming process if DSM gets filled out during interview [38]
Estimations	Minimum effort of acquisition [45]	Doubtful data

well as Dong, states that especially multiple-choice questions constrict possible responses, which can lead to information loss [28, 38]. Furthermore, Dong mentions that experts may tend to describe a target state instead of the state as-is, and then responses can hardly be justified. And Bartolomei describes the "a priori problem" as being a critical challenge [26]. This problem means that the act of information acquisition itself may constrain a system by creating inappropriate assumptions through the framing of questions.

In general, analog surveys are designed for the interrogation of many people with affordable effort. The success of such surveys highly depends on the preparation of questions and the motivation and commitment of the participants. This is because the completion of forms often cannot be controlled. Even if the effort required for conducting analog surveys is relatively low, it increases with the number of participating people and the quantity of question rounds.

Today, digital surveys are widely applied and services like SurveyMonkey (www.surveymonkey.com) are easy to apply even for first-time users. Digital surveys overcome the boundaries of conventional analog surveys. These boundaries are set by the number of interrogated people, the amount and complexity of considered data and the possibilities of automated evaluation. Unfortunately, digital surveys often seem to be even less binding for participants than analog surveys. This can result in very low response rates from participants.

6.4.2.2 Documentation

Often, structure information can be extracted from existing documents. Lindemann et al. identified several relevant sources, e.g. project management charts or the documentation of design methods like quality function deployment (QFD) or TRIZ [2]. Dong applies requirement lists [28], Avnet collects information from project management software [38] and Browning extracts elements and dependencies from organigrams [45]. If applicable, this method of information acquisition is advantageous because of few required resources [2]. But the use of outdated or unrealistic data for information acquisition can be disadvantageous [28, 45].

6.4.2.3 Observation

An observation is a methodical approach on information acquisition, where the knowledge owner is not directly occupied with the formulation of dependencies. The system modeller attends the daily workflow of experts or participates in meetings or workshops. Then the modeller creates the structure based on the insights gained. For example, Avnet took part in design sessions as a passive observer [38]. He describes that observations help increase the system understanding. For this reason, observations can be preliminary work for later interpretation of surveys and interviews [38]. And the direct integration of system modellers into the workflow of knowledge owners can enforce fast conflict solving [28]. It must be mentioned that observations can be demanding for the system modeller, and he needs to be a semi-expert. Avnet describes possible information loss as a disadvantage of observations, which can result from the organizational hierarchy and social pressure in observed meetings. And the system modeller must be aware of his own subjectivity in interpreting the observations made [38].

6.4.2.4 Interview

In many publications, interviews are mentioned as a methodical approach for information acquisition. Dong applies interviews in two case studies in the automotive industry [28, 29] and Black et al. develop a DSM describing the development process of automotive brakes [36]. Rowles applies interviews for pre- and post-processing a survey on the development of a jet engine [32]. Also Browning and Sabbaghian et al. use interviews for the preparation of a subsequent survey [40, 45]. Whereas many authors mention the application of interviews, only few methodical details are described. One of these descriptions is given by Bartolomei, who mentions the formulation of open questions for his approach on qualitative knowledge construction [26]. Intensive research has been undertaken concerning interview types and their application. For example, Kvale provides a comprehensive overview of planning, conducting, analysing and validating different types of interviews [46].

In the information acquisition of system structures, the methodical application of interviews needs to be increased. Interviews seem to enable the reliable identification of direct dependencies and the simple resolution of disagreements [28]. However, Bartolomei mentions the danger of intentional concealment by the interviewed experts to be difficult to identify [26]. And Avnet mentions the disadvantage of a time-consuming interview process if dependencies are documented during interviews [38].

6.4.2.5 Estimations

In general, system modellers can complement information, if not provided by experts or other resources during the acquisition. This can become necessary if acquiring information from the right sources would be too costly. If technical values are on hand, sometimes interpolation can be applied. If not, estimations can be useful. Browning describes a situation where knowledge owners were mandatory but did not participate in the survey. Browning estimated lacking information in order to be able to establish the system structure [45]. It has to be mentioned that reliable estimations can ask for expert knowledge. And uncertainty of the decisions made must be taken into account.

The six methods presented in the paragraphs above all possess specific advantages and disadvantages. These are aggregated in Table 6.2. For the conduction of high-quality structure acquisitions some authors combine two or even more methods. Especially interviews are often applied for validating information acquired by surveys. In these cases interviews are not used for entire system modelling, but for quality improvements only, as interviews represent the most resource-demanding method. In this context, Dong mentions that surveys result in worse results than interviews [28]. Also the use of available documentation always needs to be considered before applying other, more resource-consuming methods. In general, the specific project situation asks for the application of adequate methods. Decision parameters are mainly the system size, number of knowledge owners, available resources and required quality of resulting structures.

6.4.3　Barriers Against Successful Information Acquisition

When interacting with complex systems it is useful to be aware of the typical mistakes. Dörner explains failures people make when interacting with complex systems [47]. All of these failures can be directly related to a lack of managing a system's interdependencies. Dörner states that people often do not analyze problems adequately and therefore the concepts of problem solving are only based on small subsets of the entire system; large numbers of elements and their connectivity to the system are then neglected.

Often systems get considered as being uncoupled aggregations of subsets. One reason for that can be a person's own core of knowledge. Simply said, an electrical engineer will likely tend to search for a solution in an electrical (sub)set of the system than in a mechanical or software set. However in the case of such a one-sided problem consideration, if changes happen to non-considered system parts then the system behavior appears to be unpredictable.

Another typical mistake when dealing with complex systems is the insufficient consideration of side effects. For example, this can happen when single key figures get used for assessing complexity. As complex challenges go along with non-transparency, the desire for a simple assessment possibility becomes understandable. Unfortunately, only very specific complex system perspectives can be rated by a condensed value without neglecting significant aspects. In computer science complex problems can be differentiated by their degree of computational difficulty [1]. The required computing time or the minimal size of required computational code (Kolmogorov complexity) are used as metrics, assuming that the problem can be mathematically formulated (see Sect. 3.2). However, complex systems often contain unknown interdependencies which cannot be described by mathematical equations, and simple complexity indicators only take a small part of the system into consideration. For example, a popular system complexity rating is by its number of components; they are easy to identify and count. However, this complexity assessment does not consider interdependencies, so integrated versus modular product designs are not rated differently by this approach.

If undesired effects result from interacting with complex systems, then people tend to oversteer in order to correct these effects. The system's non-transparency, however, does often not allow one to predict the impact from these chosen measures, which can—because of manifold interdependencies—bring up further undesired effects. And it often seems to be impossible to systematically search for a solution to a complex problem, because limited resources seem to be insufficient for investigating a large-sized system. In such a situation, tendencies towards authoritarian decision making can often be seen, even if a sound decision basis is non-existent.

Unmanaged, uncontrolled complexity can result in a lack of decision-making abilities, incorrect decisions with significant negative impact, frequent changes due to lack of sustainability or long process durations. But as introduced in Sect. 6.3.3, successful management of complexity enables tremendous opportunities for enterprises. For example, a large number of product variants can be developed, maintained and offered to the market if the mutual interdependencies can be handled. Or more customer requirements can be realized when complex processes are well-controlled—which means that iterations are unlikely and change impact is predictable. The benefits of controlled complexity are so significant that many successful enterprises work with systems just at the edge of manageability.

The methods applied for acquiring system information (see Sect. 6.4.2) are designed for achieving high-quality information at manageable effort and take initial requirements (see Sect. 6.4.1) into account. Several barriers exist in application, which make system information acquisition a challenging task. Next, these barriers are introduced, because awareness of existing barriers is mandatory for future improvements in information acquisition.

Figure 6.7 indicates ten barriers to successful information acquisition and associates them with a basic product design process. These barriers have been documented by authors dealing with requirements and methods for information acquisition. In the upper part of Fig. 6.7, the result of general process steps of product design is depicted: product specification, functional structures, principal solutions, module structure and preliminary layout. Information acquisition about system structures is typically conducted while structuring the modules and creating the preliminary system layout. Knowledge owners contribute to this

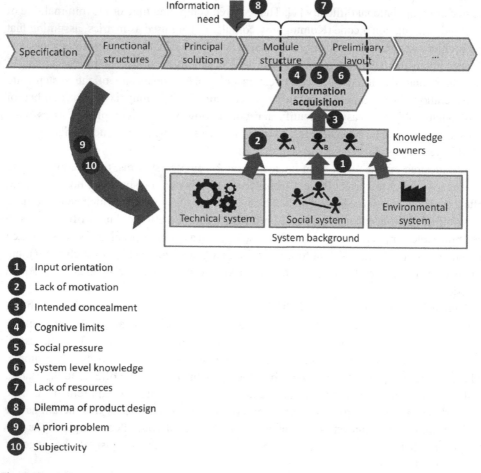

1 Input orientation
2 Lack of motivation
3 Intended concealment
4 Cognitive limits
5 Social pressure
6 System level knowledge
7 Lack of resources
8 Dilemma of product design
9 A priori problem
10 Subjectivity

Fig. 6.7 Barriers to successful structure acquisition, appearance in the development process

information acquisition using their knowledge background about the technical, social and environmental systems. The ten barriers are explained in the paragraphs below.

6.4.3.1 Input Orientation

Browning as well as Eppinger state that people's thinking is input-oriented [34, 45]. "It's important to focus on input rather than output because we have found that managers, engineers, and other product-development professionals are more accurate in identifying what they need to know than in describing what others need to know" [34]. As the input for one expert represents the output of another one, information has to be merged in order to achieve a consistent structure. The use of different wording can make such merging difficult. In Fig. 6.7, this barrier is located between the knowledge owners and their system background, as the input orientation can hardly be influenced by knowledge owners actively.

6.4.3.2 Lack of Motivation

It may happen that interrogated knowledge owners are not sufficiently motivated to provide the required system information [32, 45]. One possible reason is that the knowledge owners cannot see the benefit for themselves and may interpret the interrogation as unprofitable extra work. Furthermore, they may see the interrogation as documenting their job role concretely so they can be replaced more easily by another worker. Support from higher management will often be helpful [32]. However, while emphasizing the importance of this task, this does not solve the problem of unseen benefits. Specific interview techniques and early project involvement of experts could help increase the motivation [46].

6.4.3.3 Intentional Concealment

Bartolomei mentions the risk of intentional concealment by knowledge owners as a possibly severe barrier [26]. Especially, if knowledge owners are not aware about the usage of information provided by them, their personal interest can influence the acquisition results negatively. In this context, Dong explains that knowledge owners tend to provide desired target-states instead of as-is states [28]. Plausibility checks and the interrogation of several knowledge owners can help to identify false information. However, this barrier should be resolved at the social level.

6.4.3.4 Cognitive Limits

Three barriers could be identified, which can appear directly in the information acquisition process (see numbers 4, 5 and 6 in Fig. 6.7): cognitive limits of interrogated experts, social pressure and the need for system-level knowledge. Dong mentions that the limits of human cognitive abilities have to be considered when managing complex systems [29]. Browning relates this limitation to the handling of DSM [48]. He notes that even system models containing a few elements only can overburden people. The barrier of cognitive limits could for example be resolved by representing only the required system information and to select appropriate, case-specific visualization techniques.

6.4.3.5 Social Pressure

Avnet as well as Dong describe that social pressure, for example resulting from workshop participants' places in the company hierarchy, can lead to information loss in acquisition processes [28, 38]. The effect is that statements made by higher-ranked participants are not criticized, or lower-ranked participants do not even impart their opinion. The barrier of social pressure is closely related to the prevailing meeting culture. If necessary, workshops participants should be at the same level in the hierarchy.

6.4.3.6 System-level Knowledge

System-level knowledge is mentioned by Dong as the compilation of many people's component-level knowledge [28]. The author describes that system-level knowledge is required for system analysis, but cannot be acquired from a single expert only; rather it must be compiled by the aggregation of many experts' component-level knowledge. Thus, it is important not to simply analyze structures at the component level and directly draw conclusions at the system level.

6.4.3.7 Lack of Resources

Whereas the above mentioned barriers that can occur in the acquisition process, lack of resources can impede the general process being conducted. Avnet describes shortening interviews, because the required engineers did not have time for extensive meetings [38]. However, time or resource constraints correlate negatively with the achievable information quality. Consequently, if resources are reduced the system structure should be acquired on a less detailed level. However, too abstract of a model may result in trivial analysis results.

6.4.3.8 Dilemma of Product Design

The earlier that changes and improvements are implemented in the design process, the easier (and cheaper) their realization gets. But earlier actions mean less knowledge (and more uncertainty) about the system in question. This is a general dilemma of product design, which is also valid for the application of system structures [29]. Structure acquisition is typically done after defining general system modules (see location in Fig. 6.7). However, knowledge about basic dependencies would be helpful for determining these modules—but the knowledge is not available at that time. Techniques of early evaluation of properties could help in overcoming this barrier in the future.

6.4.3.9 A Priori Problem and Subjectivity

The barriers of a priori problem and subjectivity are related to each other. The a priori problem means that assumptions brought into the information acquisition process by system modellers can constraint the resulting system. As these assumptions are not the result of well-founded information from system experts, they can falsify the model [26]. A typical example is the constriction of a survey's solution space by wrong assumptions implemented in the questionnaire. Such assumptions often result from the subjectivity of system modellers. But not only system modellers, all participants in an information acquisition process bring in their own background [28, 32, 38]. This can impede the quality of resulting systems and can for example be counteracted by comparing information provided by several experts. In Fig. 6.7, the barriers of the a priori problem and subjectivity are depicted as impacts leading from the design process (i.e. from the daily work of experts) to the system background (i.e. experiences that shape personal knowledge).

6.5 Complexity Management Implementation

The objective of the final step within the complexity management framework is the method application to the system in order to realize the selected strategy. At this point of the process, the modelled system structures have been transferred into system knowledge; constraints and potentials of the system should be uncovered and meaningful interaction with it should be possible in order to improve the system. Improvement in this context refers to the initial questions answered during the step of system definition: What is the objective of the complexity analysis? Which results are expected? The acquired system knowledge needs to be transferred into actions for solving the complexity challenge.

The objectives are as manifold as the possible solutions and implementations are diverse. In general, the system understanding acquired in conducting the previous steps of the framework can be implemented on the complex system in two different ways. The first one aims at improving the interaction with a complex system so that complexity does not get reduced or avoided, but becomes more manageable. This can be realized by implementing possibilities of accessing, navigating and searching in the complex system structure, for example by means of adequate visualization and filtering. The second possibility of implementing the acquired system knowledge to the complex challenge is by improving the engineering system itself; this means to rearrange, eliminate or integrate new elements and dependencies within the system in order change the amount of complexity that has to be managed. For example, modularizing a so far monolithic product structure represents the implementation of this approach.

When implementing a solution one has to keep in mind that this is based on a system model, which represents an abstraction of the real system. Thus, even when the system modeling was conducted carefully, it might happen that the derived theoretical solutions do not represent viable options in practical application. Plausibility checks are mandatory before initiating the implementation.

In addition, one has to keep the importance of a holistic system view in mind when working on complex challenges. Each single measure elaborated in the selected solution approach might be easy and meaningful to implement in practice; however, different measures can influence each other when applied to the system. Thus, all measures have to be checked in their entirety for meaningfulness from a holistic system perspective.

The complexity management framework introduced in this chapter has been illustrated with examples from the field of structural complexity, because this concept is widely applied in and is the fundament of many engineering methods. But the framework also matches with other modeling approaches. The very simple six-step approach is not meant to describe the exact process of solving complex engineering challenges. These processes are highly iterative, because knowledge about the non-transparent complex system can often only be obtained gradually. Thus, assumptions made initially in the process have to be revised when new knowledge is on hand. In fact, the presented framework shall present a guideline assuring that important steps towards the solution of a complex problem are not skipped and are done in the right sequence. While this seems to be obvious, practice shows that it is often not respected. The reason for that can be found in the typical failures made when interacting with complex systems (see Sect. 3.3), as they have been described by Dörner [47]. So it can happen that one selects a method for solving a complex problem even before an appropriate system definition is obtained. A person might be familiar with a specific method and the situation seems to require immediate action—thus well-known tools often become the first choice. For sure, this is not a promising approach. One has to be aware at all times that dealing with a complex engineering challenge means to be confronted with a decision based on incomplete system knowledge. So a stepwise, systematic clarification of causes, effects and objectives is required for reliably determining appropriate strategies, methods and implementations.

References

1. Mainzer, Klaus. 2008. *Komplexität*. Paderborn: Wilhelm Fink.
2. Lindemann, Udo, Maik Maurer, and Thomas Braun. 2009. *Structural Complexity Management— An Approach for the Field of Product Design*. Berlin: Springer http://medcontent.metapress.com/ index/A65RM03P4874243N.pdf.
3. Browning, T.R. 2001. Applying the Design Structure Matrix to System Decomposition and Integration Problems: A Review and New Directions. *IEEE Transactions on Engineering Management* 48(3): 292–306. doi:10.1109/17.946528.
4. Maurer, Maik S. 2007. *Strcutural Awareness in Complex Product Design*. Munich: Dr. Hut. http:// nbn-resolving.de/urn/resolver.pl?urn:nbn:de:bvb:91-diss-20070618-622288-1-1.
5. Senge, Peter M. 1994. *The Fifth Discipline: The Art and Practice of the Learning Organization*. New York: Doubleday.
6. Sterman, John D 2000. On an Approach to Techniques for the Analysis of the Structure of Large Systems of Equations. *SIAM Review* 4(4).
7. Kusiak, Andrew. 1999. *Engineering Design—Products, Processes and System*. San Diego: Academic Press.
8. Melnikov, O., V. Sarvanov, R. Tyshkevich, and V. Yemelichev. 1994. *Lectures on Graph Theory*. Mannheim: BI-Wissenschaftsverlag.
9. Schuh, Günther, and Urs Schwenk. 2001. *Produktkomplexität Managen. Strategien, Methoden, Tools*. München: Hanser Fachbuch.
10. Kreimeyer, Matthias F. 2009. A Structural Measurement System for Engineering Design Processes.
11. Ashkenas, Ron. 2013. *Simply Effective: How to Cut Through Complexity in Your Organization and Get Things Done*. Harward Business Press.

12. Anderson, Chris. 2008. *The Long Tail: Why the Future of Business is Selling Less of More*. Revised ed. New York: Hyperion.
13. Wildemann, H. 2008. *Komplexitätsmanagement in Vertrieb, Beschaffung, Produkt, Entwicklung Und Produktion*. 9th ed. München: Komplexitätsmanagement in Vertrieb, Beschaffung, Produkt, Entwicklung und Produktion.
14. Jania, T. 2004. Änderungsmanagement Auf Basis Eines Integrierten Prozess- Und Produktions-Modells Mit Dem Ziel Einer Durchgängigen Komplexitätsbewertung. Universität Paderborn.
15. Wikipedia.org. 2015. List of Nokia Products. https://en.wikipedia.org/wiki/List_of_Nokia_products#Mobile_phones.
16. Battista, G., P. Eades, R. Tamassia, and I.G. Tollis. 1999. *Graph Drawing: Algorithms for the Visualization of Graphs*. Upper Saddle River, NJ: Prentice Hall.
17. Lima, Manuel. 2011. *Visual Complexity—Mapping Patterns of Information*. New York: Princeton Architectural Press.
18. LaValle, Steve, Eric Lesser, Rebecca Shockley, Michael Hopkins, and Nina Kruschwitz. 2010. *Big Data, Analytics and the Path from Insights to Value*. MITSloan Management Review.
19. Eppinger, Steven D., and Tyson R. Browning. 2012. *Design Structure Matrix Methods and Applications*. Cambridge, MA: MIT Press.
20. Gießmann, Marco. 2010. Komplexitätsmanagement in Der Logistik. Kausalanalytische Untersuchung Zum Einfluss Der Beschaffungskomplexität Auf Den Logistikerfolg. BoD—Books on Demand.
21. Gebala, David A., and Steven D. Eppinger. 1991. Methods for Analyzing Design Procedures. In Proceedings of the ASME Third International Conference in Design Theory and Methodology, 227–233. Miami: ASME.
22. Baldwin, C.Y., and K.B. Clark. 2000. *Design Rules—The Power of Modularity*. Vol. 1. Cambridge, MA: MIT Press.
23. Felfernig, Alexander, Lothar Hotz, Claire Bagley, and Juha Tiihonen. 2014. *Knowledge-Based Configuration: From Research to Business Cases*. Waltham, MA: Morgan Kaufmann.
24. Daenzer, W.F., and F. Huber. 1999. *Systems Engineering: Methodik Und Praxis*. 10th ed. Zürich: Verl. Industrielle Organisation.
25. Hub, H. 1994. *Ganzheitliches Denken Im Management: Komplexe Aufgaben PC-gestützt lösen*. Wiesbaden: Gabler.
26. Bartolomei, J. 2007. Qualitative Knowledge Construction for Engineering Systems: Extending the Design Structure Matrix Methodology in Scope and Perspective. Engineering Systems Division. http://hdl.handle.net/1721.1/43855.
27. Alexander, C. 1964. *Notes on the Synthesis of Form*. Cambridge: Harvard University Press.
28. Dong, Qi. 1999. Representing Information Flow and Management in Product Design Using the Design Structure Matrix.
29. ———. 2002. Predicting and Managing System Interactions at Early Phase of the Product Development Process. Department of Mechanical Engineering, no. February 22:19–296.
30. Eichinger, Markus, Maik Maurer, Udo Pulm, and Udo Lindemann. 2006. Extending Design Structure Matrices and Domain Mapping Matrices by Multiple Design Structure Matrices. In Proceedings of the 8th Biennial Conference on Engineering Systems Design and Analysis (ASME-ESDA06). Torino: CD-ROM.
31. Biedermann, Wieland, Matthias Kreimeyer, and Udo Lindemann. 2009. Measurement System to Improve Data Acquisition Workshops. In 11th International Design Structure Matrix Conference, ed. Matthias Kreimeyer, Jonathan Maier, George Fadel, and Udo Lindemann. Greenville.
32. Rowles, C.M. 1999. System Integration Analysis of a Large Commercial Aircraft Engine. Masters Thesis in Engineering and Management. http://dspace.mit.edu/bitstream/handle/1721.1/9753/42769852.pdf?sequence=1.

33. Ahmadi, Reza, Thomas A. Roemer, and Robert H. Wang. 2001. Structuring Product Development Processes. *European Journal of Operational Research* 130(3): 539–558. doi:10.1016/S0377-2217(99)00412-9.
34. Eppinger, Steven D. 2001. Innovation at the Speed of Information. *Harvard Business Review* 79 (1): 149–158.
35. Eppinger, Steven D., and Vesa Salminen. 2001. Patterns of Product Development Interactions. In International Conference on Engineering Design ICED 01. Glasgow.
36. Black, Thomas A., Charles H. Fine, and Emanuel M. Sachs. 1990. A Method for Systems Design Using Precedence Relationships: An Application to Automotive Brake Systems.
37. Pimmler, Thomas U., and Steven D. Eppinger. 1994. Integration Analysis of Product Decompositions, no. September.
38. Avnet, Mark S. 2009. *Socio-Cognitive Analysis of Engineering System Design: Shared Knowledge, Process, and Product*. doi: 10.1016/j.anpedi.2008.08.013.
39. Glaser, Barney, and Anselm Strauss. 1967. *The Discovery of Grounded Theory*. Chicago: Aldine.
40. Sabbaghian, N., S. Eppinger, and E. Murman. 1998. Product Development Process Capture and Display Using Web-Based Technologies. *IEEE International Conference on Systems Man and Cybernetics* 3: 2664–2669. doi:10.1109/ICSMC.1998.725062.
41. Biedermann, Wieland, Ben Strelkow, Florian Klar, Udo Lindemann, and Michael F. Zaeh. 2010. Reducing Data Acquisition Effort by Hierarchical System Modelling. In Proceedings of the 12th International DSM Conference, ed. Matthias Kreimeyer and David Wynn. Cambridge, UK: Hanser.
42. Guivarch, Antoine. 2002. Car Engine Design Process and Organization Management Using DSMs. In 4th Design Structure Matrix Workshop, ed. Steven D. Eppinger, Daniel Whitney, and Ali Yassine. Cambridge, UK.
43. Sauser, Brian. 2006. Toward Mission Assurance: A Framework for Systems Engineering Management. *Systems Engineering* 9(3): 213–227.
44. Björnfot, A., and L. Stehn. 2007. A Design Structural Matrix Approach Displaying Structural and Assembly Requirements in Construction: A Timber Case Study. *Journal of Engineering Design* 18(2): 113–124.
45. Browning, Tyson R. 1996. Systematic IPT Integration in Lean Development Programs.
46. Kvale, S. 2008. *Doing Interviews*. London: Sage.
47. Dörner, Dietrich. 1997. *The Logic of Failure: Recognizing and Avoiding Error in Complex Situations*. Cambridge: Basic Books.
48. Browning, Tyson R. 2001. Applying the Design Structure Matrix to System Decomposition and Integration Problems: A Review and New Directions. *IEEE Transactions on Engineering Management* 48(3): 292–306.

Summary and Future Challenges

7

This thesis presented an introduction to complexity management in engineering design. Complexity often seems to be a fashion term, which does not only appear in our working environment, but also pervades many aspects of private life. Everybody talks about the ongoing increase in complexity. And especially engineered systems seem to participate in this increase faster than average, hence managing complexity becomes an important competency for engineers. For this reason, the thesis provides a comprehensive picture of engineering complexity from different perspectives.

While the term complexity seems to be omnipresent, there is no common understanding about its distinct definition, even if focused on the field of engineering only. However, it appears there is a general agreement that complexity in engineering is harmful, therefore undesirable, and consequentially is something that should be avoided. And it seems that we are not too successful in our attempts to avoid complexity, as a popular platitude says that "the world is becoming increasingly complex". And everybody experiences the increase in complexity in daily life.

The beginning of this thesis provided some relevant basics about complexity in the context of engineering. First, the general composition of complex systems is explained, followed by a distinction between the terms complicated and complex. While these two terms are mostly used synonymously in everyday language, for practical applications this distinction is extremely important, as approaches for solving complicated versus complex problems are significantly different. An overview of complexity definitions relevant to the engineering domain shows the scope aggregated in this single term and emphasizes the need for a detailed understanding of the type of complexity in a specific use case. If one definition of complexity helps to rate the required change effort for a software project, the same definition does not necessarily serve well for reorganizing the engineering team that

© Springer-Verlag GmbH Germany 2017
M. Maurer, *Complexity Management in Engineering Design – a Primer*,
DOI 10.1007/978-3-662-53448-9_7

develops a service-intensive mechatronic system. The introduction of relevant definitions is followed by a basic introduction of four approaches towards complexity which are commonly applied in engineering practice. Of course, practice in engineering development will yield a broader knowledge of approaches, models and methods for dealing with complex challenges in engineering. The overview given in this thesis only introduced some of the most popular approaches and provided a general understanding of typical complexity handling in engineering development. With the background of general engineering complexity, next a historic perspective of the emergence of system thinking and complexity was presented.

The historic developments in system thinking, system theory and cybernetics are highly interwoven. And the presentation in this thesis cannot even lay out the influences and trends provided by all relevant contributors from different sciences and arts. In fact, in the past many influential artists were also acting as scientists. And for example criticism at the reductionistic system approach did not only come from scientific standpoints, but also from an artistic movement like romanticism. In this thesis the presentation of the scientific step provided the basis for understanding modern approaches of complexity management. It is interesting to see how closely system thinking is related to other scientific findings and discoveries. And how new findings and technical challenges exposed older approaches as insufficient, and so prepared the path to developing improved models. Considering the entire development of system thinking, it is noticeable how popular the reductionistic system model was until recent times. And while the simplicity and clarity of a reductionistic system model is still useful for many applications, one should be aware of its limits—and that alternative approaches exist, which are applicable beyond these limits.

The presentation of historic developments was succeeded by a categorization of recent research fields working on complexity and its management. This perspective on state-of-the-art research helps to comprehend the current focus on complexity management in engineering. The definition of categories follows its reference in literature. The partly diffuse application of terms and differences in vocabulary would also allow for different assignments of topics to the categories shown. And when adapting the level of detail, even additional categories could be implemented to the picture in this thesis. Because of the broad scope of complexity issues, it was not surprising that many overlaps could be identified between the research categories. And future transfer and integration of scientific approaches can fill currently existing blank spots. In such a large and multilayered field like complexity management, it is not possible to create a generally accepted depiction of the subject matter. Each perspective results in a different perception and interpretation of the approaches towards complexity management. The purpose of the depiction in this thesis is to motivate the discussion about the state of the art and the future relevance of topics. The selected Venn diagram is easy to access and adapt, and should serve as a starting point for future discussions.

Even if the illustration and description of specialized topics in complexity management in this thesis is comprehensive and easy to access, this perspective does not allow a direct application to complex problems. Chapter 6 described a framework to guide practitioners through the relevant steps in solving a complex engineering problem. The framework is generally applicable and not exclusively linked to a specific modeling concept or scope. The application of the framework is introduced by focusing on structural complexity and using dependency modeling. The reason for focusing on these aspects of structural complexity is the significance of successful complexity management approaches like modularization. In this framework, special focus is given to a thorough system definition, which must precede any determination of strategy and method application. This avoids the frequent mistake that one selects a strategy and method only because one is familiar with them—and not because they meet the requirements for solving the specific challenge. And in the step of system modeling, the framework strengthens the importance of information acquisition. Especially in the engineering domain, system information like elements and dependencies need to be acquired from knowledge owners, and the acquisition can be highly complex. The requirements of, methods of and barriers to information acquisition for complexity management were presented in detail. An interesting fact is that information acquisition is not intensely covered in literature. This step in complexity management is decisive for the model quality and consequently all further interpretations, conclusions and measures based on the system model acquired. Being aware of the requirements, methods and barriers to information acquisition presented in this thesis can help increase the model quality significantly.

Part of this introduction to engineering complexity was the explanation of the historic development from early system thinking to recent complexity management techniques. The historical view highlighted that interacting with systems and their inherent complexity did not represent a recently emerging scope of duties. In fact, complexity challenges and approaches towards their understanding, analyzing and managing have been subject to continuous evolution for more than two millennia, having chosen Aristotle's approaches to system thinking as the starting point. In the past, discoveries like Isaac Newton's classic mechanics were the decisive foundations for reductionistic and deductive system modeling approaches. And the pressure to innovate in the Second World War led to the framework of cybernetics, the first modern trans-disciplinary approach towards managing complex systems, whereby systems could comprise machine and human components for the first time. There is no indication that the further development of system modeling and complexity management will decelerate. Quite the contrary, since technical progress advances increasingly fast and innovations with worldwide impact seem to occur in shorter time intervals. The world gets more and more networked and with it the recent global challenges become more complex. With a background of system thinking it is unthinkable that challenges like global warming or worldwide refugee movements can be solved by only local or national approaches.

Recent times not only possess specific complexity challenges, but also yield advanced approaches, methods and techniques for their management. Chapter 5 provided an overview of current topics in complexity management research from an engineering perspective. Unsurprisingly, the large main topic can be classified into several subsets as specialization continuously increases. While several overlaps between disciplines exist and the increase of knowledge may be a natural reason for developing more branches, this specialization can implicate a significant risk. The historic development has shown that complexity management and the underlying systems thinking are strongly based on interdisciplinary approaches. So if the specialization of disciplines dealing with complexity results in compartmentalization with difficult mutual exchange, then the new upcoming challenges of complexity can hardly be tackled successfully. Increasing differences in applied vocabulary can be one of the rising barriers that work against fruitful cooperation between disciplines. Thus the tremendous increase of knowledge seems to make specialization inevitable, and focusing on specific aspects even supports further partial developments in complexity management. But solving the challenges of the future, which represent complex systems of systems, requires intra- and transdisciplinary approaches. Therefore, special attention has to be paid to creating overlaps and exchange between specialized domains.

Complexity seems to be required for realizing higher states of development. Goldberg and Holland mention that "the 'genetic programs' of even the simplest living organisms are more complex than the most intricate human designs" [1]. And several authors conclude that engineering design needs to be inspired by biological systems. Vester formulated this in his 8th rule of biocybernetics. In this rule he claims that products, functions and organizations "must conform to the structure of viable systems". Vester describes that "non-biological design ultimately fails to address the relevant demand and as such is produced without regard to the market. Yet countless planning disasters continue to result from decision-making processes that ignore this rule" [2].

Following the ideas of bio-inspired system design and considering that biological systems have reached states of robust complexity with impressive functionality and flexibility, one can argue that future engineering challenges will not only be about reducing, avoiding and controlling complexity, which results from engineering tasks as (mostly) undesired byproducts. That would mean that the intentional design of complexity for targeted engineering of powerful functionalities is the future of complexity management. Such functionalities would go beyond the possibilities that can be realized by existing engineering approaches, and could for example be located in fields like autonomous systems and artificial intelligence. Of course, the creation of such new engineering opportunities will require intra- and transdisciplinary collaboration, as it was already successfully applied to system and complexity science in the past. The continuously increasing speed of knowledge generation and increasing fragmentation of disciplines impose high demands, which have to be met in future research and application. The next logical step is to embrace complexity as a design element.

References

1. Goldberg, D.E., and John H. Holland. 1988. Genetic Algorithms and Machine Learning, 95–99.
2. Vester, Frederic. 2007. *The Art of Interconnected Thinking—Ideas and Tools for Tackling Complexity*. Munich: Mcb Verlag.

Index

© Springer-Verlag GmbH Germany 2017
M. Maurer, *Complexity Management in Engineering Design – a Primer*,
DOI 10.1007/978-3-662-53448-9

Printed in the United States
By Bookmasters